# MEANWHILE FARM

Margaret Cheney

# MEANWHILE FARM

LES FEMMES
MILLBRAE, CALIFORNIA

*To my mother, Jo Swisher,*
*whose students are legion.*

ACKNOWLEDGMENTS

I am grateful to Harcourt Brace Jovanovich, Inc. for permission to reprint excerpts from A ROOM OF ONE'S OWN by Virginia Woolf, copyright 1929 by Harcourt Brace Jovanovich, Inc.; renewed, 1957, by Leonard Woolf; and to *Mother Earth News* for permission to quote excerpts from the Ploughboy Interview with Steve Baer, Issue No. 22.

M.C.

**CELESTIAL ARTS**
231 Adrian Road
Millbrae, California 94030

First printing: September 1975
1 2 3 4 5 6 7 8 / 80 79 78 77 76 75
Made in the United States of America

Library of Congress Cataloging in Publication Data

Cheney, Margaret.
    The Meanwhile Farm.

    1. Farm life--California.    2. Cheney, Margaret.    I. Title.
S521.5.C2C45    979.4    75-9442
ISBN 0-89087-905-2

# Contents

. . . They too, the patriarchs, the professors, had endless difficulties, terrible drawbacks to contend with. Their education had been in some ways as faulty as my own. It had bred in them defects as great. True, they had money and power, but only at the cost of harbouring in their breasts an eagle, a vulture, for ever tearing the liver out and plucking at the lungs—the instinct for possession, the rage for acquisition which drives them to desire other people's fields and goods perpetually; to make frontiers and flags; battleships and poison gas; to offer up their own lives and their children's lives. Walk through the Admiralty Arch . . . or any other avenue given up to trophies and cannon, and reflect upon the kind of glory celebrated there. Or watch in the spring sunshine the stockbroker and the great barrister going indoors to make money and more money when it is a fact that five hundred pounds a year will keep one alive in the sunshine. . . . Indeed my aunt's legacy unveiled the sky to me, and substituted for the large and imposing figure of a gentleman, which Milton recommended for my perpetual adoration, a view of the open sky. . . .

—Virginia Woolf
*A Room of One's Own*

# Chapter I

# Swamp of the Hawk

"Some places in this canyon, they've got a water shortage," the real estate agent had said. "I guarantee you won't have one here."

I put on boots and crept through the barbed wire fence to wade across the field. The weeds towered above my head, except in places where they had been crushed flat by sleeping deer.

Aside from the underlying earthquake fault and the wetness, there were only a few drawbacks. The Benevolent Cement Company, our leading polluter, had a right-of-way through the middle of the place for hauling limestone from its quarry in the nearby foothills. There was as yet no water well, utility line, or road through the bog. No Welcome Wagon. But the sycamore tree, like a gigantic canary cage, was exploding with birdsong on that spring morning.

Five acres of swamp, meadow, and wooded hillside set in a canyon that managed to seem both untamed and friendly. At sunset when the trees that fringed the hills to the southwest were sharply illuminated, it would look wild and even sinister. The Old Stage Road that had once provided the main land link between San Diego and San Francisco still followed the profile of those hills, plunging down through country so dense and harsh that you could believe in the living presences of Black Bart, Murietta, Three-Fingered Jack, and the extinct California grizzly bear.

A certain curve on the perilous grade was known as Hat Corner. Old-timers recalled it was at this point, as a rule, where the stovepipe hats of the dudes and nobs from the East would be snatched off by the wind and hurled down the incline.

From the southwest and Monterey Bay the fog rolled in most evenings, up over the hills, fingering in through the canyons and across the flat, lush valleys. The lowlands, hot by day, produced row crops, fruits, and nuts for the supermarkets of America. In the rolling foothills, thousands of acres of ranchland had undergone transformation to new vineyards. More change lay ahead, in the name of progress.

This entire county, bypassed and hidden-seeming, still had a population of little more than thirteen thousand. But the freeways were being widened all the time despite fuel shortages and auto cutbacks. Smog, no respecter of boundary lines, was rolling over the hills from urban counties to the north. Each year more of the beautiful orchards were reduced to stump fields, the stumps bulldozed and burned. Yet vast areas of the county remained open and children still grew up here, able to see how wild creatures lived in their natural surroundings.

I glanced at my watch. Still time to get back to the real estate office before the broker went home.

I found him waiting confidently.

"How much?" I asked.

"Only three thousand an acre. Little enough when you consider those artesian springs. But cash is scarce right now. If you can pay cash, I'll give you fifty percent off. Prime agricultural land around here is going for five."

"Uh. What other kinds of costs would I have?"

Obligingly he ran up a tape on his desk calculator. The figures he handed me provided estimates on road-building, well-drilling, a septic tank, drainage, and the construction of a small dwelling—the bare beginning costs of a farm, without even a chickenhouse or a packet of seeds. At the same time I was toting up my meager resources in my head. The gap between his tape and the one in my head seemed too large. One decision I had made was not to go into debt. Although I had been earning a more than adequate living I had also managed in the most patriotic tradition to spend most of my salary by the end of each month.

Did I really need or want a farm? Why this? Why now?

The sixties had begun for most of us as a time of idealism and renewed hope. But political assassinations, burning cities, tear gas and bullets on university campuses, and the dragging on of war had killed just about everything.

Midway in that chaotic decade my husband had died tragically. When the seventies came I could not seem to let go of the sixties.

My only child went off to college. Then a favorite aunt died, leaving me a bequest. The years had spun by and I seemed perpetually trying to catch at some clue in the swiftly receding distance, some fragment that might promise a renewed continuity to life. Back there in the dark woods I—and I believe many of my compatriots—had dropped the torch. Now we had to stumble around and find it and see if a trail existed that led out of here to somewhere—maybe in time all the way to the place that Winston Churchill once spoke of as "the sunlit uplands."

Was it fair or reasonable to want a farm simply because one had not managed to make peace with history? But if you failed to salvage something from the past, slay some of its demons, how could you hope to confront a future?

A quiet place was needed—and time, time. Yes. I really needed and wanted this particular farm. But I did not choose to be a loner. And this could be no ordinary farm. It needed an extraordinary cofarmer— or rather *I* did—who felt about the important things as I felt and who had been through approximately as much as I had been. Someone who was compatibly irreverent of sacred bulls and could recognize a time for happiness somewhere short of perfection and who, above all, possessed that marvelous, strengthening, blessed quality—a sense of the absurd. These thoughts were groping for form in my mind as the real estate broker waited for my decision.

"I have a friend," I said, "who might be interested in buying the place with me. She once lived in this county. I know she loves the area. Probably, though, she would not be able to live here for a year or so. I'll talk with her and let you know."

Barbara Nelson was the one person I knew who might possibly be flexible and foolhardy enough to enjoy such a soggy Arcadia. I took her to see the swamp and the hills. She was enchanted and said that in perhaps a year she might be able to arrange an early retirement from her job in a nearby city.

When we again called on the broker to initiate the purchase, he congratulated us on having found "just the ideal size of a farm." He was right about that all right. How were we to foresee that, before the first pear trees had borne fruit, the same gentleman would be leading a realtors' attack on the concept of the five-acre spread? (How, that is, unless we were in the habit of reading the daily papers?) They would say it was too small to farm and too big to keep neat—as if anybody ever moved to the country out of a passion for neatness—while ardently touting new zoning for cluster developments with big cookie-cutter condominiums for the beautiful, narrow canyon.

We *did* move to the country out of a passion for serenity, not expecting to become involved in the hottest land-war to flare up in that county since sheepherders and cattlemen had bushwhacked each other in the range battles of the nineteenth century.

That day in his office, though, we were all great conservationists on the same team.

"The Benevolent Cement Company," he said, "is in violation of the environmental protection laws." The gritty old smokestacks that powdered much of the canyon with a thick gray dust would have to be modernized or closed down. What Barbara and I saw that day as an ideal for the future of the region, after an end to local pollution, was the gradual development of new farmsteads and the growth of clean, small industries within commuting distance.

Barbara started calling the place *Meanwhile*, which captured its feeling, and ours about it. The name was also a ritual nod toward the transiency of life in general and a bone for the implacable resident jinn at the bottom of our earthquake fault.

Our lives, like those of thousands of women, had once been irrevocably changed by a reading of Virginia Woolf's essay, "A Room of One's Own," and we could not overlook the fact that an aunt's bequest had figured significantly in the acquiring of this farm as in her "room." Women generally had been denied by the dominant culture a tradition of time and privacy unrelated to biological/patriarchal supporting roles. For most women that was still true. The all-important "habit of freedom," as Virginia Woolf put it, essential to the development of women as writers, poets, thinkers, artists, and politicians remained a hard financial thing to achieve.

As Woolf said better than anyone has since, a woman must first free herself from, "All desire to protest, to preach, to proclaim an injury, to pay off a score, to make the world the witness of some hardship or grievance . . ." Only then could that special "incandescent" state of mind denied to Shakespeare's sister and to his sister's sisters all down through history, hope to burn as a universal beacon. We were still far from having attained that beatitude.

Neither my cofarmer nor I aspired to scrawl, "Sister, Shakespeare's" on our country mailbox. A farm called Meanwhile felt to us like a place not apart from the world, but a source of strength within it from which useful work of other kinds might flow. On the positive side, it seemed to us that the seventies had ushered in the glimmering of a Golden Age for women in the arts and letters.

Men, of course, also dreamed and schemed of Meanwhiles. Life was grievously hard for *anyone* with a revolutionary idea, whether in poetry, prose, paint, stone, or papier mache—whether thinker or tinkerer.

Men too sought margins of freedom in which to develop their personal work. Untold numbers of *their* lives ended violently and prematurely, depriving *our* lives of richness. For industrial society exacts a heavy toll of the creative person. Beyond that, creativity itself demands that one lay everything on the line, risk all that most people value—power, security, money, status, sometimes even love.

There are Meanwhiles and Meanwhiles. This book is about the building of only one kind, which I suppose should be called the low-budget Arcadia.

Probably there are as many kinds of Meanwhiles as human beings. I have known people to develop good Meanwhiles without budging from their own homes in great cities. For some, the Meanwhile is a mere continuum of a former career. An emeritus professor of journalism I know goes each day to teach parttime, there being nothing finer to him than the company of his students and colleagues. I know a librarian who all during his working career went to auction sales and bought Victorian furniture and brought it home, until his house contained sixteen upright pianos, three baby grands, two harpsichords, and four hundred and thirty-one chairs. Somehow his marriage lasted. Then the man retired and went to work on his Meanwhile—restoring the furniture lovingly, piece by piece.

I doubt that there is any ideal age for a person to begin a Meanwhile. One takes the plunge when one is ready to confront such sacrifices as it may demand. It is advisable to keep one's eyes open during this type of plunging. But if you know what you really want, no price you are prepared to pay nor any effort is too great.

After I moved to the canyon I visited Meanwhile farms that other amateurs had had simmering on the back burners of their dreams for quite a while. I saw them grow and flourish. I saw each small farm taking on a character distinctively its own. The successes far outweighed the few that did not pan out. One reason for this is that there are few production-line lags on a Meanwhile. The Meanwhiler works like a demon and loves it.

Where our Meanwhile was to rise from the marsh there were sycamore, willow, and cottonwood trees that traced the bed of an iron-brown creek. The canyon floor was no more than a quarter of a mile wide at this point. Perfect for hurling echoes, should one's special talent lie in that direction.

Beyond the canyon north and south were overlapping contours of hills, green in the contrasting shades of oak, madrone, horse chestnut, maple, and chaparral. In spring when the poison oak leaves are shiny-green and virulent, they would complement the misty clouds of blue ceanothus, the wild lilac of the Western coastal woods.

I noticed at once the special windswept purity of the canyon sky. In late afternoon the strong western wind, smashing into warm interior air from the agricultural valleys, created dramatic cloud effects. Redtail hawks and turkey vultures skimmed down from the highest peak to glide across the meadows in swift pursuit of their prey. They would hitch rides aloft on the thermal current, screaming with delight. On the highest range of hills, on the peak itself, the air often seemed to blow straight up from the valleys. If you were camping there, it could lift your tent and bounce it down the slope.

The heat, the hardness of the soil, the intense blue of the sky, all reminded me of an inland Greece. I would think often, in the canyon, of how it must be to scratch a living on the rocky islands of the Aegean Sea.

And the weeds! Never had I seen them in such variety, so strange and vigorous, so many that pricked, burned, blistered, poisoned, or clung. I had never seen so many herbs that were fragrant and useful, so many that promised consolation if not cure. At least five kinds of thistle grew robustly in the meadow. One variety resembled a medieval bonker that might have been used to subdue knights-in-armor. It had the head of a giant sunflower, but with spikes instead of petals, and some called it Russian thistle. Poison hemlock grew there too and horse nettles as tall as trees, as well as blackberry vine, wild mustard, burdock, foxtails, sawgrass, cord grass, spearmint, velvet grass. There was also the jointed swamp weed that Indians were reputed to have used for scrubbing their pots, and burs in fiendish abundance. Aeons of trial and error had gone into the evolution of the swamp growth until everything that had managed to make it at all made it superbly.

Later, when I was planting seeds, I would be impressed repeatedly by nature's way of snatching things out of my hands and doing it to suit herself. For example, the seeds I would plant in a certain plot might fail to germinate, but one or two blown by the wind or carried by water to some better location would catch on and suddenly spring up with amazing vigor. A tropical plant that I had brought from Berkeley, set out in the canyon, pampered and crooned over, soon died. Yet later I was puzzled to find the same kind of plant growing lustily in various parts of the yard. After a bit of detective work I traced it to some wild birdseed that had been blown from a feeder. If nature gave something the okay, you could scarcely root it out with a crowbar. But if I sowed the seed, each day would become a small battle for survival.

Where the canyon floor was not actually under water, the soil tended to be heavy and dark, glistening with an oily, mica-like substance. In places there were extensive gluey deposits of adobe clay as well as streaks of blue potting clay—scarcely what the valley farmers would have called "prime agricultural land." A layer of hardpan about a foot below the surface offered the most challenging drainage problem any fool could ask for.

The swamp also provided a lively and robust colony of fauna, including black widow spiders, mosquitoes of generous proportion, mealybugs, scale, ants, gigantic black beetles, earwigs, inch-worms, knot-knee nematodes, gopher snakes, racers, rattlesnakes, king snakes, mud turtles, tarantulas, centipedes, field mice, domestic mice, kangaroo rats, water rats, rat-rats, moles, voles, shrews—and thousands of pocket gophers whose tunnels and mounds were everywhere. (What pocket gophers keep in their pockets is the telephone numbers of more pocket gophers.)

I could have believed anything about this bog—that it was what the UC Extension experts called pear decline country; maybe avocado root-rot country; or that poisonous gas flares burned at night. In my wildest dreams I could not have imagined a place more beautiful, more desirable. Nor could my dog Alice. She promptly located a shiny brown "potato bug" two inches long and began pushing it around with her nose. Later I asked the county agent about this bug, which was as large as a tarantula and moved in a peculiarly invincible way, like a juggernaut. He told me it was a prionus and that, once its wings grew, it could fly like a bullet, more bird than bug.

The small creek that paralleled the hill flowed thinly and muddily but otherwise unpolluted out of limestone (it was silted-up by the run-off from the quarry stripmining). It splashed down ten feet or so in a waterfall whose music would soon be audible from my bedroom window. In winter it would become a roaring cataract.

At night, as I soon discovered, all manner of hairraising creatures gave tongue—katydids, frogs, owls, coyotes, wild dogs, wildcats, bobcats, an occasional mountain lion, and monster black Angus range cattle, all up there in the dead of pitchblack earthquake country, bawling their hearts out in lust and loneliness.

Equally disturbing were the phenomena of full moonlight and star brilliance. They cast an eerie day-glow on the meadow and edged the black contours of the hills. The effect was oddly like that of a nocturnal seascape. I half expected to see a small sailboat gliding across the silvery swamp.

How long had it been since I had breathed air so pure, seen stars so fiery, felt silence so encompassing? And how much longer, given the inroads of man, could even this small miracle of wasteland survive?

After closing the deal, I asked a longtime resident, "What does *cienéga* mean? I keep seeing the word on the little property map."

"Swamp. When I was a child we always spoke of this part of the canyon as the little swamp."

Later I learned that the canyon and the hills for miles around had all been part of an enormous Spanish land grant known as the Rancho Ciénega del Gabilan. The Swamp of the Hawk.

## Chapter II

# "We Were Always
# an Eyesore"

The beauty of this fen was its unpromisingness and its uncompromis-
ingness. No danger of a woman going soft or getting the smarthead.
There was also enough of the hard, unpleasant, non-potential, enough
grubbiness and anti-chic, to discourage for a while at least the wrong
kind of city folk. Also, Fester would like it here. Fester was a plant
that—or rather, *who*—ate bugs. His only requirement for perfect happiness was one fly every two weeks. He seldom felt upwardly mobile. On
the other hand, you could always tell if he was languishing because his
pot got moldy. The day we moved to the ciénega, it seemed to me that
he perked up.

Waking up is the finest part of a country day, stealing a few minutes
to watch the sun move across the trees and clumps of chaparral on the
crest of the hill, slowly enveloping the world in splendor.

In midmorning the farm dog Alice leads the way to the mailbox, a
ritual passage through fields to a link with a distant planet and an
unreal past. The mailman usually toots if he finds anything interesting
in the daily batch.

Today is a three-toot day.

Alice prances and bucks, proud to be carrying the best dead mole in
her spring collection. On the return she will fetch a Coors beer can
collected in the wake of tourists bound for the state park. Although I
was glad to have her engaged in good works, it was hard to get rid of
recyclables.

When she had first shown an inclination to collect every beer can in
sight, I had praised her; but I soon learned to be cautious, for Alice is
very achievement-oriented. In no time at all she was ranging far afield,
even bringing me much of a farm neighbor's garbage—the baby's pants,

old socks, old gloves, the empty baby food bottle, the duck carcass.

Once, at wit's end for something to bring and to show-and-tell, this devoted animal even brought me the feed bucket of the neighbor's horse, which she had dragged home through several fields and under barbed wire without once spilling ten pounds of mixed oats and barley.

This morning, after bringing in the beer can, she would have to turn right around and go back to retrieve her mole. Being in the right place at the right time, *with the right materials*, DOING THE RIGHT THING, could be tiring. Even as a pup, she often sighed and groaned in her sleep.

The mole would be returned to her treasure heap to rest safely beside the worn skeleton of her acorn woodpecker, her unusually rich lump of horse manure and, among other choice items, a sailor cap named Ricky in gold letters—for she was nothing if not discriminating. Her treasures, whether funky or classic, pop, or plain rotten, would always be the finest examples in their categories. Best of Litter.

Letters today—from nineteen-year-old daughter Amanda Drop-out, selling her blood and beads in Athens to help finance world travels and the formation of a multinational import-export corporation. From nephew Robert who builds domes. Perhaps it was his letter that had elicited the toots from the Postman. Addressed to "Mother Meanwhile and Her Chickens," it bore in the space where a return address would normally appear a warning, "The Earth is Cracking, Cracking!"

Other letters—from old friend Anna Yee, M.D., who had retired early (but now, it seemed, not early enough) for travelling and who wrote that she was consulting a psychiatrist who specialized in ministering to patients with terminal illnesses. From Terry, a young editor and poet recently moved to New York. I walked slowly back through the field, reading:

"This city is making a true feminist out of me. Am working with an antirape group to repeal the idiotic corroboration requirement from state law: 'In order to facilitate conviction, would you mind watching while this nice man rapes me?' 30,000 rapes in the city last year, 30 convictions. Everything is really makebelieve, they are singing, singing. Don't you believe it. I got ripped off again a couple of weeks ago and soaked for new locks on my sometime residence to the tune of $100 . . . and at work, a discreet note from big boss saying I had to wear shoes at *all* times . . ."

Ah, world!

Opening the morning paper, I found myself drawn helplessly to the letters columns in which people bared souls and other parts. A friend of mine had noted sadly that just as you were getting to know someone through Dr. Wartspore's column, you lost him. Thirty-two years old . . . the left testicle undescended . . . I whisked on to Dr. Frank Miller's Wonderful World of Animals.

"Dear Dr. Miller: Our canary is a good singer but for some reason he seems to hate me. My wife and daughter can come right up to him and it doesn't bother him a bit. But let me approach within five feet and he goes into a rage. He makes *angry peeping noises* (my italics) and stretches out his wings and puts his head down as if he would tear me apart. How can I make friends?"

I was certain Miller himself invented these rending appeals, an excuse to reply: "The canary may hate you (sort of), considering you a rival—for territory or mate. Conversely, he may be strangely attracted to you. His behavior may actually be a courting display, not really meant to discourage you at all."

Over the years, on the theory that everyone enjoys reading other people's mail, I have measurably reduced the time it takes me to reply to my friends' letters by sending them letters from other correspondents (after a proper introduction, of course). Now I thoughtfully scissored out the letter about the canary that wanted to rape the man's wife and daughter, if not the correspondent himself, with the idea of forwarding it to Terry. Perhaps, after considering the matter, I might decide to send the pertinent section of her letter to Dr. Miller, with the request that he forward it to the bird. I enjoyed putting people and others with similar interests in touch. In honesty, I had also lost a few friends.

I let the newspaper fall to the floor, and then for the first time noticed a cheap, small envelope among the advertisements. The address was childishly begun with red pencil and finished off with green ballpoint-ink. Obviously the pencil lead had broken and the green ink looked as if it had been on the verge of running out. It—the envelope—looked the way envelopes often do if an infant has been hiccoughing near them. I turned it over and noted with pleasure that it was from the Oleanders, my friends of the New Whole World Revolution. They had lived in the canyon and built fences for me; but destiny summoned and they were now settling on land at Garberville near the Mendocino county coastline.    Marcie wrote:

"Quiet moments, Heather sleeping, dreaming baby dreams, bread baking in the oven, puppies playing on the grass, yellow, purple, white flowers swaying in the breeze, chickens pecking at the bugs happy to be in sun.

"On April 28 at 8:13 p.m. our baby girl was born. I was in labor for 21 hours at home and we decided we should go to the hospital to make sure everything was OK. It was, just a long slow labor. Anyway it all turned out well. My parents were especially happy we went to the hospital. Tom got to be with me while Heather was being born, so next time he'll know that much more.

"About the bird cage, the chickens are still using it at night. We haven't built a chicken house for them but we will soon. We will have to 'cause they won't fit in the cage for too much longer. We really appreciate the use of it."

(Another cage I had lent the Oleanders for their cats when they moved from the canyon had been left at my front door, slightly mashed, with a note: "We are sorry that the truck rolled on it.")

Marcie's letter continued: "How is your garden doing? We have one started, the soil here seems very good once we add some organic matter to it. It is very dark which is good but gets very hard packed.

"Baby is waking up so I will say good-bye."

Her letter closed with the benediction from The Incredible String Band, that was also engraved on the Oleander's hand-built camper: "May the long time sun shine upon you/ All love surround you/ And the pure light within you guide your way on." (Tom Oleander had explained that it might keep the farmers from shooting at them.)

Twenty-one years of age. In labor for twenty-one hours. And already planning a second child for home-delivery. With pioneers like the Oleanders, assuming they did not get carried away with procreation just for the sake of obstetrical practice, it was hard to see how the Revolution could fail. Or, for that matter, how the AMA could survive.

Oleander was of course their assumed name. Keeping the one you were born with, like submitting to being branded with a social security number, was in Tom's view "What the *Bible* calls the mark o' the beast." I found myself wishing I had told them the name I had favored for their baby—Bangladesh Oleander.

Meanwhile was the first of the small, new farms in the canyon. Most of my neighbors were in the process of starting their families—former city people who worked full or parttime at outside jobs, schoolteachers and such. All of us slightly dazed and humbled at the good fortune that had guided us to where we were; sometimes nervous or, speaking for myself, plain scared. There was the night the lights went out, shortly after I had moved there. Even the moon and the stars went out. The telephone rang in that blackness. A shaky voice said, "Are you all right?" The caller was my only neighbor at that time.

"Fine, fine," I replied with scarcely a break in my voice. Only after hanging up did I remember that she was pregnant and that her husband worked nights in a distant city. Well, maybe it had been a comfort to her to know that I could fumble the telephone off the hook in an emergency.

You got so you could almost tell which farmer was going to stay the course by whether he started out making a big splash with the genetic purity of his livestock. I am not sure what the connection is, but this farmer's wife would soon begin to miss her Tupperware parties and babysitting services and before you knew it they'd sold off the expensive breeding stock and moved back to town. The experienced ranchers carefully crossbred their animals. One nearby farmer had imported wild-eyed Scottish Highlands cattle which he was adventurously crossing with the pink-and-white Charolais to achieve hardiness with lowfat tenderness.

I pondered what I might do with Fester, my adventure in botanical pest-control (One Step Beyond, as we called him). And then I decided I was not going to be stampeded into any darnedfool risk-taking.

The Meanwhile farmhouse, a monument to zealous amateurism, had begun to rise—not on the hillside but down where the marshlife could be observed. And I made sure it was surrounded by a no-nonsense tangle of barbed wire fencing. Here Fester and I, and the rest of our world-weary group could close our eyes, take a deep breath, and conjure up the romantic atmosphere of Mann's *Death in Venice*, complete with sinking feeling. Fester's expression began to burn with a hectic, fevered light.

Among the kindred spirits who turned up were a young couple on the far side of the nearest hill, both of whom were of the new breed of snake-loving schoolteachers. They told me they had moved to the country to escape from the emotional stresses, not of violent city streets, but of a tiny farm in a suburb where a developer had promised they would find "total security, with all the eggs you can eat."

They had wound up with two hens brooding at the apex of a nasty, precarious pyramid of forty-seven eggs. Since every new egg was immediately added to the tower, the young marrieds never knew which was fresh and which foul and had been forced to buy their eating eggs at a market. The cost of chicken feed plus fresh eggs, they said, brought on a financial crisis. The constant anxiety of not knowing when the whole mess might collapse and inundate their tiny house produced a combination of terror that drove them back to the real estate person. I fully understood their present plan to grow only a few vegetables and wildflowers on their acreage.

I offered them a clipping of Fester.

"Maybe later," the husband said quickly. "Right now we're into rattlesnakes."

When another neighbor shot a rattlesnake on their place, they claimed they spent a whole month getting it replaced.

Soon I was visited by Amy, another newcomer. She was a potter. Her husband worked in a nearby town and they planned to grow vegetables organically for a market route. I started expounding my favorite theme about how a farm, in comparison to a trimmed green square of suburbia, ought always to look quite messy. (Head 'em off at the gap.)

"And the reason for this," I declared with the authority of three months on the land, "is that you are always working on jobs-in-progress. The seasons are always changing. The job you started in summer probably won't be *quite* finished when autumn comes. So there you are with the old tools and odd bits of material, when you have to dig out the new tools and new kinds of material, and nothing quite ready for the finishing touches. But it's just *Nature's way.* Universal rhythm system; and no use to fight it."

"I understand exactly what you mean," she said, as I thought a trifle too hastily. "When Angelo and I lived in Los Angeles, we had this house that had been trapped between two freeways. The paint, that he described as tit-pink, was drifting off in flakes. We were surrounded by weeds that, once in a while, we managed to half-cut. But we were already planning our escape to the farm, as soon as there was a gap in the traffic long enough for us to get the car onto a freeway. Well, we finally made it. Yes, we were *always* an eyesore."

"Good girl!" I said. "If any noseys from a neighborhood improvement society come around, send them to me. What I'll tell them is that I'm a recycling depot. They can't very well knock that, can they, what with Alice's little scavenger service?"

When I first settled in, a well-disposed longtime resident (few longtimers *are* well-disposed toward shorttimers, and a farmer who had been around for fifteen years told me they still called *him* a newcomer) advised me, "Just lay back and see how things go for a while. Most people make the mistake of rushing into buying animals. They also plant too much. You can spend a fortune planting all the wrong things the first year. Get acquainted with the weather, your soil, and your markets."

It was the most valuable farming advice I have had. Even today I still plant a lot of wrong things and will go on doing so, but only when I can afford to experiment, and knowing it is purely that. (My secret addiction is to seed catalogs from all over the country.) So I tried to control the convulsive jerking of my green thumbs. There was much else to do that summer just in the race to provide shelter. Still, I couldn't help coveting the beautiful lambs and calves and ducklings and piglets in other people's pastures.

Finally I took the bit in my teeth, tossed my bonnet over the millrace and my pink plastic curlers into the gooseberry bush. I ordered a dozen Bantam chicks from Clarence at the feedstore. Life was a gamble—wasn't it?

Clarence said there was a fifty-fifty chance of some coming in around the middle of summer.

"That's good enough for me," I said. "I couldn't ask for better odds. And you won't catch me spinning my wheels either. How is your hayfever?"

"Worse," he said.

"You ought to change jobs, Clarence. You really ought. You look just terrible. Imagine working in a feedstore with hayfever."

"Hay is my life," he replied sadly but with dignity.

When we new farm folk lied to each other, the *in* topics were: permanent pasture (didn't matter that you had no livestock, you could still talk a good line of legumes and green manure); irrigating with a windmill or a portable gas pump in your own pond; hatching eggs with the heat of methane gas generated by chicken droppings (which came first, we asked, the chicken or the gas?)—any fuel left over from the incubator to be used to run our pickup trucks. We talked about building all-year even-temperature root cellars to keep our roots in; about raising our own beef for our own freezer, which would be powered with electricity from the water wheel in our creek, that would also turn the stones that ground the wheat and corn. If we were stark-raving mad we talked about making the farm pay for itself inside a year. If we spoke of making a slight profit the first year, only *mild* insanity was indicated, but insanity nonetheless.

I soon learned that, with fairly simple old-fashioned gamesmanship, I could establish a good deal of authority among the other tyros. If some new dude said you only had to grease your windmill once a year, I spoke up smoothly and quickly with, "Yes, but not in the North." And it made little difference that *none* of us owned a windmill, yet.

By memorizing the names of weeds in the fine print on a bottle of herbicide, I acquired some small reputation as a botanist. And no one seemed to notice that the names of species I was rattling off were more applicable to the American South, East, and Midwest than to our canyon. Then Amy came along, with her esoteric knowledge of herbal medicines and teas, and started smoking red raspberry leaves hand-rolled in cigarette papers—creating fumes that smelled remarkably like grass—and introduced us to the high you can get from Valerian tea; all of which detracted somewhat from my own fragile reputation.

People kept looking at the swamp and asking me what my cash crop was, even though it looked to me as if anyone could see I was just laying back, analyzing my markets and tax shelters.

They would say something like, Is it geese?

I would fold my arms, lean back in my director's chair and say, Nope. But that wouldn't stop them. They would say, Artichokes? And I would say, Jerusalem or regulars? and they would say, Regulars, and I would say Nope, and then they would say Apricots? and I would say, Blenheim or Burgess? and they would say Blenheim and I would say Nope. Maybe then they would say, Clover-willow honey? and I'd say Nope. Sage honey? Nope. Fresh-herb wine vinegar? Nope. And I would add, Can't you see that clover doesn't grow in this swamp, it gets *root-rot*, man; and they would say, Well then is it catfish? Nope. Comice pears? Nope. Bloodworms? Nope. Chenille bedspreads? Nope.

Finally I started saying the cash crop was watercress, because that grew wild and very well. I could as easily have said sphagnum moss, since I had a bogful of it, but I could not bear to think of the questions that might have led to. Then, as luck would have it, drought struck. The swamp steamed and shriveled up, the cash crop went to seed, and the farm began to look like Death Valley in miniature, warmed over. Even so, people kept on unfeelingly asking about my cash crop. They wouldn't take gila monsters for an answer either, even though it was reasonable enough.

Luckily the whole cash crop picture changed once again, after my talk with Clarence. I went legit. From then on, when the curious asked, I could answer, Banties. I figured the ball was in Clarence's court now.

All this, however, was prologue: I had not even tested the soil at Meanwhile to find out what if any crops could be grown—a little matter that smart farmers thought to look into before ever closing the deal.

Chapter III

# Great Moments in Modern Farming

As the Great Goddess Isis discovered in her early horticultural research, all plants have a pH preference. A pH of 6.5, for example, denotes a range of acid to alkaline content that is desirable. Soil testing kits may be bought, and a farmer can make her own analysis. I felt, however, that I would not believe the results unless they came from the horse's mouth, as it were. There was also the likelihood that even if I knew what my pH factor was, I still would not know what to do with or about it. Whether to rejoice or lament, that would be the Q. So I called the county agricultural agent's office and asked to speak to the Goddess. Great was my discomfiture and confusion on learning that the only female on the staff was a clerk-typist doubling as receptionist, who also had complete responsibility for Extension Service leaflets.

However, she introduced me to the county agent's assistant, a recent A&M graduate, who soon called on me at the swamp, equipped with tools for core sampling.

He examined a few tubes of compacted dirt and said, "You want the truth, don't you?"

"Not necessarily," I replied intuitively.

He told me that at a foot below the surface almost everywhere (on the *dry* side of the field), he had struck water floating on top of hard-pan. The latter is a natural cement-like substance that usually runs an inch or so in depth. With great effort, one could hack and bore down through the hardpan in each planting hole to create drainage so that small trees and vines could spread their roots and, possibly, survive. For every plant, my informant said, this would mean arduous labor, with no assurance that root-rot would not ensue anyhow.

"What do you think I should do?" I asked.

"Have you thought of moving back to the city?"

"Not a chance."

"Well," he said, "you might want to try growing what we call companion crops. Say, weeping willows and watercress—."

"I wonder if there is much of a market for bogmallow?"

"No, Ma'am."

Casting around for a more cheerful subject, I said, "Do you happen to know what that tree is near the fenceline, the one with the dull-green leaves and the yellow-orange tubular flowers? It looks like a wild honeysuckle."

"Yes, Ma'am. That is South American tree tobacco. It is deadly poisonous to man and beast and it has recently begun to spread rapidly through parts of North America.

"And that lush green growth over there," he added enthusiastically, "is horse nettle. If you touch it, it will raise a blister immediately—even right through your jeans. You know the poison oak of course?"

I nodded.

"And this poison hemlock, which looks a great deal like wild carrot but is not? If you boil the leaves, you can make the same tea that Socrates perished of."

I had surreptitiously brushed a horse nettle with one finger. A blister sprang up so quickly that I slipped my hand into my pocket.

He whacked a few samples of a kind of grass with soft tassles, saying, "I'm not sure what *this* is, but I'll look it up when I get back to the office."

Later he wrote me a letter which, first of all, verified a dismal pH range. Just 100 percent dismal.

He added, "The tall grass I brought to the office to identify looks very much like velvet grass or *Holcus lanatus*."

Reading that, I brightened a little. It was like the time the orthodontist said, "You have no occlusion whatever," and I ran home yelling, "Look, Ma, no occlusion!"

I continued reading: "*Holcus lanatus*, or velvet grass, is common in California and grows in damp areas. It does not make good forage."

With the letter he had enclosed a leaflet on the commercial growing of watercress. It contained some good tips on controlling leeches.

Once in a while when I need a lift, I drive a short distance to visit some friends on the Monterey Peninsula who are expert birdwatchers and almost always have what is, to me, cheering or diverting news of recent traffic on the Western Flyway. Florence and Ed Wilson have a small business and they keep *Peterson's* and *Birds of North America* handy in the office so that arguments can be settled quickly. They always know exactly which migratory birds are in town or at what pond in the suburbs and whether the Christmas count of shorebirds was up or down, and whether the yellow-billed loon could be seen at the Great Tidal Pool.

After we had exchanged our respective bits of news: "A rosebreasted grosbeak at Crespi Pond. Imagine! And downtown Pacific Grove is *full* of killdeer. They almost *never* come into populated areas." "And I had soil samples taken at my new farm and guess what—*Holcus lanatus!*" After we had been through that, we settled an argument about a tiny brown bird I had seen at Meanwhile. Florence quickly identified it as a hermit thrush, finding its picture for me. Ed spoke up and asked if I knew about the condor.

"What condor?"

"The California condor, of course. One of them was seen in your canyon just a while back."

"Fantastic! You can't mean it?"

Like all but a few people in the world, I had never seen a California condor. It is closer than anything alive today that might possibly be compared to our jolly old extinct friend the pterodactyl or flying serpent. The condor has evolved beyond the reptile stage, but it is huge, it is largely a glider, and it has long been on the Endangered Species list. Fifty or sixty remaining California condors live in a remote preserve in the Los Padres National Forest near Santa Barbara. They are awkward and rightly edgy birds that like to sit around with hunched shoulders. Their wingspread is sometimes more than ten feet. A few more condors have been sighted living in the mountains of Mexico. The names, ranks, and serial numbers of these survivors of the species are well-known to the Audubon Society. Everything possible is done to protect them from disturbance.

Every once in a while a forest fire breaks out in the Los Padres range. For a few weeks we condor-worshippers hold our breath.

And that is why I was so incredulous at the news that one had been sighted in the canyon.

"You could probably be appointed by the Audubon Society as a condor station for your area," Ed said. "That is, if you wanted to."

"If I *wanted* to!"

But then, thinking it over, I stammered, "N-no: the p-pressures. The way my luck runs, if I ever sighted a condor there would be an earthquake at the same time. The sky would go dark from a rush of great wings, the earth would rumble and split open at my feet, thousands would fall into the chasm and die. The prospect of becoming a condor station is tempting; but no. I just can't risk it."

I said goodbye to the Wilsons and set off for my new home. As I reached the canyon, the usual redtailed hawks could be seen gliding overhead. I almost wrecked the car craning my neck out of the window, which brought me to my senses. So what if I *didn't* become a condor station? So what if the swamp farm did have a crumby pH factor of minus 0.031? There was still a good chance that some morning, going out to feed the Banties—when I *got* some Banties—I might look up at the sky, and suddenly—there it would *be*!

Anyhow, there was nothing to stop me rushing out at dawn every morning to give the sky a checking-over; plenty of farmers did that. For far smaller reasons.

# Chapter IV

# Witching the Well

The most urgent job was the drilling of a well. No water, no farm. A swamp, to my astonishment, provided no guaranty of a supply of potable water, or at least so the driller told me.

I was excited about the well. Pure water had become so rare, both pollution and the doctored flavor of city water so commonplace, that I could not even imagine what well water might taste like. Just the idea of connecting with an underground stream as it flowed toward the sea through strata of geological ages assumed the nature of a religious experience. Human being that I was, if a thing seemed mysterious, it proved not that I was ignorant but rather that God must have a more than usually direct hand in it. So Alice and I began to visualize a kind of god inhabiting the bottom of a deep, cool well, looking something like a pale, wise, omnipotent frog who perhaps spoke in tongues.

I was astonished when the driller hired a well-witcher who turned out to be a perfectly ordinary-looking man.

Dowsing or witching is a standard method in this area for locating a drilling site. The formal name of the game in times past was *rhabdomancy*. But today's new interest in expanding the limits of consciousness has also extended the scientific lexicon and, under the general heading of extrasensory perception (ESP), we find *radiesthesia or dowsing*. Impressive labels of this sort attest to man's unremitting effort to imbue mystery with scientism. It is perhaps not too cynical to suggest that they also help to persuade federal agencies to shell out research grants. In any event, the witcher with his willow wand is trying to tune in on hygrometric influence. Like the intrauterine device, anaesthesia, hypnosis, chatting with plants, and so on, it works for some people but not for others. No one can clearly explain why.

The important thing is that it usually works—even though the U.S.

government is somewhat schizoid about the method. For example, governmental agencies, while not officially recommending it in publications, regularly and successfully use well-witching both at home and in overseas water-searching projects. Universities, ever leary of the off-beat, do not as yet confer a doctorate in rhabdomancy, which seems a pity.

My local water driller applied his knowledge of surface and subsurface geology but backed it up with the dowser's vibrations. If there had been a conflict between the two methods, I gathered from what I saw that they would have chosen the dowser's method as more reliable. Neither method was foolproof.

The ability to witch wells, as the gender of the word suggests, is often possessed by females. The only practicing well-witcher I met in the canyon proper was a woman. She happened also to be a former park ranger, a botanist, and an outstanding farmer. When we talked about the subject she said, "I don't like to witch wells for other people. It's such a big responsibility."

The driller's man, however, had no qualms for he made his living in this way. He looked around my farm, and said, "So you're the victim?" Warning me to watch out for rattlers, he then whittled a willow wand and began slowly walking to and fro across the meadow, which was the drier side of the field.

Dowsing for water in a swamp presented problems, for he got vibrations all over the place. He insisted, however, that some were more promising than others. He confided to me another weakness of the system.

"There are about eighteen minerals up in those hills that will give you the same vibrations as water."

I said it was all right with me if he struck gold.

I had been walking around with my own willow wand, getting no messages whatever.

The dowser passed me with his willow wand vibrating and twanging as if all the demons in Hell were signalling. I felt eaten with envy. It occurred to me to wonder whether I could tune in on another person's vibrations, so I touched his willow wand—promptly breaking the circuit. He looked at me in a hurt way. It seemed to be like jamming a radio signal. When I removed my hand he began to get signals again. I could not help feeling a failure.

"Sometimes," the well-witcher said kindly, "it works better if you use a dry willow wand. Some people can do it with a coat hanger."

He stopped, turned, and poised at a certain point as if listening as well as feeling, then continued slowly across the field.

"Best thing," he added, "is to cut and dry a willow wand and let it mellow until you get the feel of it being as much a part of you as your arms."

The woman dowser I mentioned and her husband operated one of the most beautiful farms I had seen. It was not run in the way of agri-business. There was no factory-like production where the amount of profit became the total goal and the determinant of every move. Each part of their operation reinforced another in natural harmony. Sheep grazed under the apricot trees, sheep fertilized trees, trees shaded sheep, fertilizer grew grass for sheep with the help of irrigation water for trees. The earthworms and soil bacteria were very much alive. No-body panicked and ran for the poison spray if they saw a bad bug or a gopher. To their credit it had always been operated that way. Even before Rachel Carson's time, ladybugs and lambs and every growing thing worked together. Each had its job.

This is not to say that even the wise farmer's course was always sweet-ness and light. Sometimes a late, heavy, spring frost killed the blossoms on the apricot trees. Sometimes the trees were covered with ripening fruit and, just at the last minute, an unseasonable hot spell would cause it all to drop prematurely to the ground. Apricot growers said they were never sure they had a crop until the money was in the bank, and it was the same for most people who had orchards. "A successful farmer," they said, "is one who can smile while losing money." That seemed to explain why, in the country, you saw so many happy faces.

The woman farmer who witched wells had located two excellent ones on their place, in one case disagreeing completely with the driller as to the best location. She insisted he drill where she felt the best vibrations and he brought in a flow of fifty gallons per minute.

After my well-witcher had settled on a likely spot for drilling, a der-rick was moved into the swamp and they got down to business. Several days passed impatiently for me. Finally the driller announced that he had hit, at about seventy feet, a stream producing ten gallons per minute. It was a poor performance for the area, being adequate for household use but not for much irrigation. Underground streams could change, though, particularly in that area where the earth was con-stantly being shifted by minor quakes. The source of a well might be diverted by some convulsion of the earth, but more appealingly, the flow could be enlarged by the diversion of other waters into it.

I scooped up a glass of water, still dark with the sand and mud in it, and drank. It was cold and quenching and as pure as I had hoped. But it was also hard, like most of the local water, and would sometimes be brown with iron sediment.

Because wells were expensive and the drilling only the beginning of costs, I thought a reference book ought to be printed for the benefit of new farmers. It ought to tell one how wells, pumps, and pressure tanks worked. A driller often enjoyed a monopoly in his county, making the customer completely dependent on his esoteric knowledge and his honesty in reporting how deeply he actually drilled. That much power could be tempting. It seemed to me the U.S. Government Printing Office ought to publish something for us innocents.

The Meanwhile well, electric pump, and water softening system cost almost as much as a new car. Like a new car, it represented a lot of humming, sometimes whining, snivelling, downbreaking, and generally complex technology that left the farmer at the mercy of the expert. The trouble with such experts was the same one confronting automobile owners—they were expensive, hard to get hold of when needed, and sometimes knew little more about the problem than the nonexpert.

After all the equipment was installed at Meanwhile I sometimes wished that I had one of the old-fashioned, squeaking hand pumps such as I remembered from the farms of my childhood. With these new-fangled, submersible electric motors you would have to dive to the bottom of the well if you hoped to fix one with a piece of baling wire. Still, memory tends to gloss the inconveniences. Usually I just felt grateful when I drew a glass of cold, clear water from the tap at the kitchen sink.

One of the things that had begun to irritate me was that suddenly various kinds of ugly, glaring hardware were starting to encroach between me and nature. First there had been the well driller's big pressure tank of stainless steel, which held reserves and forced water through the pipes, and which was about as tall as the farmhouse was to be. The driller insisted it had to stand upright to function properly.

Soon it was joined at the kitchen door by the gas man's lurid blue Propane tank, which had to lie on its side to perform its functions.

I said, "Don't you have a tank painted earth-brown?"

Shocked and hurt, he said he painted all his tanks Italian blue, as they called it in these parts, but that I could repaint mine any color I wished. This proved to be a reasonable offer, since I later learned I would be paying a substantial use charge for the tank for the rest of my life.

He added, "But in my opinion, there's nothing much purtier in the world than a blue Propane gas tank."

I hooked my thumbs into the bib of my denim apron, gave him my decision-making look, and said nothing. As soon as he got off the property, I was going to have a hogwire fence built around the tank farm, which I would use as a trellis for flowering vines, and the enclosure for a dog kennel. And I would paint both tanks a pleasant mudtone.

One of the things bothering me was plain humiliation at having little idea how the big tanks worked, what the various gauges and red warning devices were for.

One day I called the well driller's helper to come out and adjust the water pressure. This earnest-looking young man turned a dial slightly and gave me a bill for about the same amount a brain surgeon would charge for a transplant.

"How in heck does this thing work anyhow?" I asked innocently. "What controls the pump at the bottom? Where does the air come from that seems to fluctuate inside the tank? Why does the pump start up at the bottom of the well every time I flush the toilet? Why does the pressure gauge go up and down? How is it that—what is the relationship, I mean to say, between the pump down there and the tank up here? For example, I understand that eventually the pressure tank will get water-logged and will have to be drained. Which of these spigots would I open or close?"

He stared at me, his mouth working silently.

"You see," I continued, "I'll probably want to jerk out the casing sometime. So it would be good to know the principles."

All the color drained from his face. He continued to stare openmouthed. Then silently he began edging toward his pickup truck. And in a flash he was behind the wheel, grinding the starter.

"Don't forget your tools," I called.

In retrospect, it seemed that a twenty-foot well dug by hand among the artesian springs in the swamp might have produced excellent drinking water in adequate supply, and perhaps indefinitely. But because I was still suffering from my old middle-class hang-ups about money, I feared the water would be no good unless I paid a lot for it, which was what the well drillers relied on.

With this first step out of the way, I could proceed to other matters.

Farming was a highly individual undertaking. No one embarked on it without the itch to make changes, to impose something of one's unique personality—hence the tenacious myth of the hostile environment with its need for "control," and for all kinds of "management." Resisting the urge to enhance nature might be the hardest thing of all. My feeling was that title to the land conveyed nothing more than a leasehold, like the title to life itself. Everywhere on Earth, almost, one could see examples of how people had confused legal title with ownership and the latter with license. On the other hand, if necessary, I would resist usurpers to the death.

My rule-of-green-thumb would be, "the less change the better."

Certainly that was also the most convenient philosophy. The digging would be hard in the sunbaked land that was not actually submerged. The danger of my over-modifying looked remote.

# Chapter V

# Entwhistle's Upper Hand

With trepidation, I went to see a man named Entwhistle about the major land project. He was a big earth mover. This may sound contradictory, in light of my earlier comments; but every swamp-farm needed a pond, if only because nothing in the world was much purtier than one or more interesting than pond life.

One could not on a farm economically do everything one's self. As in urban living, it was simply a matter of trading off time. So one sometimes wound up waiting for Entwhistle.

You could not bribe such artisans for they were all millionaires. If you asked them questions, they would not answer you directly—either because millionaires didn't have to or because they had concluded you would not understand. They looked you straight in the eye and on through it, and out the back of your head, to a focal point on yonder hill. If they answered questions, they were not the questions you had asked. And finally you would begin to wonder which of you was deaf or daft.

It would cross your mind to wonder, if it is they who are daft—which you preferred to think—then how was it that *they* all drove Cadillacs or Lincoln Continentals?

I ran Entwhistle to earth driving his big tractor on a perpendicular hillside in front of a new house. He waved and made three or four more turns before he could bear to turn off the motor.

Sensing that his time was valuable, I went briskly to the point. "What are the chances of you bringing your backhoe, RD-8 tractor, steamshovel, crane, dump truck, erector set, the whole works, over to my little *ciénega?* I have a problem with the land engorging careless automobiles, which has not helped me win popularity contests. And I have a mosquito problem. They are not sleeping-sickness mosquitoes, they are alert, healthy, and active. I would like you to make a driveway and a cesspool, dig a ditch, and a nice, deep pond. How soon would you be able to get to it and how much would it cost?"

Entwhistle's clear, innocent, blue eyes looked down upon me reflectively. After a very long time he said, "You don't want too big a pond."

I squirmed up onto the hood of my car, the better to engage his line of vision.

"How soon could you get to the job?"

"All you can do is what you can do," he burst out. "I never make promises. I learned a long time ago not to make promises. Nope! Nosirree-bob! Why, they been waitin' for me two months over to the school. I told 'em the land was too wet, no use my going over there 'til it dired out. Yep."

Sly little dimples played at the corners of his mouth. He climbed down from the tractor and leaned against it, the better to disengage his line of vision.

"Yep," he repeated, as if underscoring an inexorable truth.

I waited for him to take up at least one of the questions I had asked. But the expression in his eyes suggested a person who has departed this planet, with its petty and annoying mortal problems. I slid off the hood of my car and started pivoting in the mud on one of my boot heels.

The sly dimples were coming and going more rapidly than ever and his true-blue eyes positively twinkled. I waited, scarcely daring to breathe.

"Yep," he blurted. "And they been waitin' for me over to the Grange for three months."

After that, nothing broke the silence for at least a full minute. Entwhistle, you slippery so-and-so, I thought, I can wait you out. But it was I who yielded.

"How much would it cost for the work I want? Altogether? Approximately? In rough, round figures?" (And don't let me flap you.)

Prolonged silence.

No sign that he might have heard. He was now gazing across the field and along the horizon. I saw the reflection of a computer (one of the big ones) humming smoothly away in the depths of his clear, honest eyes. An airplane passed overhead and he followed its course with such intentness that I began to wonder if maybe he had to go build a new airport. I remained stubbornly quiet.

"Yessir," he said, finally. "I learned a long time ago never to make promises. I always tell my boys, *don't make promises!*"

"And very wise," I concurred. (Since you obviously don't keep them.)

By then I was convinced that the work I had mentioned would all cost hopelessly more than I could afford. I also began to feel as if I didn't even *look* like a person who could afford that kind of work; maybe like a woman who went around engaging heavy equipment under false pretenses just to experience the surge of power. Or maybe Entwhistle had been tricked by hundreds of women who ordered his heavy equipment and then, when he showed up—ENTRAPMENT OF DUMPTRUCK OWNER BARED.

"Well," I said, "surely you can give some idea of costs?"

Entwhistle: "All you can do is what you can do. I tell my boys that. I say—. The backhoe will only reach twelve feet. That means, if you go in from both sides, you can only have a pond that's twenty-four feet in diameter, see?"

"Yes, but after you have made the twenty-four-foot pond, couldn't you go around the bank again and again, scooping out twelve feet more each time until you made the biggest pond in the world?"

He gave me a hurt look.

Encouraged, I went on. "You see, I'd like to run a little resort out there—as a cash crop. Water skiing and piranha fishing, something of that sort. Maybe for Swinging Singles."

He had begun to be impressed, I could see that, for the guile was now clear in his limpid blue eyes.

"She'll silt up on you! What you want to do is build your dam about five feet below where the opening is into the pond. Then the silt collects down in that area and she won't silt up on you. That way your pond stays clean."

"Yes," I said quickly. "About how much would that cost?"

Entwhistle: "You ought to put some gravel in the bottom of the ditch. That would help. I could bring a load of gravel on my dumptruck. Need gravel for the driveway too."

"Is gravel very expensive?"

His voice rose passionately. "Yep! Gravel is what you need all right."

"Would you say gravel costs over or under one million dollars, Mr. Entwhistle? Approximately?"

In a major concession he replied, "I have to buy it by the yard."

"How much does it cost by the yard?"

"The truck holds twenty-six yards."

We were getting warmer, so I held my silence.

"Twenty-six yards," he said, "would just about do the job. Yep. That might just about do 'er."

"Well, fine, then. I'll expect you the first of next week."

Entwhistle smiled.

"Tuesday?" I asked.

He chuckled. "I don't make promises. Learned a long time ago . . ."

"Could you just give me some *estimate* of how much the whole project might cost?" I asked, hoping to throw him off balance with a switch of subject.

Instead it plunged him into renewed silence. I began to think that Entwhistle might never speak again.

Then: "I have to charge $28 an hour for the tractor. That's scale."

I bided my time. It didn't work. Then I said, "Well, would it take one day or six days or how many days do you just *estimate* it would take (at eight hours a day at $28 an hour)? And how about the backhoe, the lifter, the truck, and the—."

"You don't want too big a pond," Entwhistle said.

"Well. See you the first of next week."

Again he smiled. He broke into chuckles.

"You are amused, Mr. Entwhistle? Perhaps you assume that I cannot afford your services?"

"I'm going fishing next week. For three weeks. Got a little cabin up in the Sierra. Yep. Go fishing up there in the mountains about this time every year. For around three weeks. Take my boys."

I could visualize them all sitting around the fire in the evening, practicing not making promises.

What I said was, "Well, I'll tell you what *I* would do if I drove something that I charged $28 an hour for. I wouldn't go fishing. I would go to Europe and possibly to the Orient, and maybe around the world. By the way—I've been thinking a little more about the pond. I think we'll plant a few trout. Rainbow trout for personal, very special friends."

Entwhistle, whistling through strong teeth: "Learned a long time ago not to make . . ."

"That Lincoln Continental over there," I said, "is that yours?"

Slowly, cagily, he admitted it was.

"I'm surprised," I said. "It looks tarnished. I thought you'd probably drive a new one."

". . . promises."

I left. I could see the crack about his car had gotten to him. Just let him sweat.

Many an exciting horticultural breakthrough has been made by small farmers waiting frustratedly for the big equipment men.

Ruth Stout, the fiery New England organic gardener, hit on her brilliant idea of year-round mulch while pacing around waiting for a man with a plow. In so doing, she succeeded in dispensing forever not only with the plow but with fertilizing, composting, weeding, spraying, most watering, and indeed almost all the labor of farming except trotting in from the garden with armloads of freshly picked vegetables. What could possibly be wrong with that picture?

And when, I wondered, would children be able to read of her achievements in their textbooks, see her face on a coin or a postage stamp? When would her figure appear in a statue at the municipal fountain, Ruth-Stout-beating-plowshares-into-mulch? When? A-ha! But perhaps you will have noticed that men are still racing around with plows and disks attached to tractors? Sore losers? (We shall go into the mystery shrouding the Stout method a little later on.)

During the agonizing weeks when I was forced to wait for Entwhistle, I fully expected some barrier-crashing technological discovery to appear to me in a vision, especially since I was hallucinating quite a lot just then—without chemical or other aid—mainly about idyllic farms that ran themselves, had two of every kind of animal who had their young in the springtime, yet never became overpopulated so that you had to kill your intimates; and about the farm being so prosperous that the farmer travelled widely every fall to study foreign methods. But no breakthrough came in this time of fever. Time passed—.

Pride kept me from telephoning the unspeakable Entwhistle.

During this period, Fester the bug-eating plant took ill. After a dramatic seizure, and despite the application of heroic measures, he was pronounced dead by an acupuncturist of my acquaintance. We entombed him with his favorite female wasp.

And then on a fine and frosty morning, just when I had about given up hope for the pond, there arose such a clatter out on the county road that I flew to a window to see what was the matter. And there it all was, trying to back and fill and squeeze through a normal-sized gate—tractor, steamshovel, and a green flatbed truck about two blocks long, loaded with one or more smaller tractors. The roof of the cab had an airhorn befitting an Alfa Romeo. The engine billowed diesel fumes from a tall, silvered smokestack. A very small man, Young Entwhistle, sat at the wheel in the cab.

Somehow this whole circus was soon inside the field, led by Entwhistle the Elder in his Lincoln Continental. I drew him into the kitchen with coffee. His son, he told me, was an artist with the tractor.

In a moment, all the equipment was rushing back and forth through the swamp, churning, and grunching. Alice barked. The diesel fumes were thick as tear gas. Mr. Entwhistle and his son were looking very happy. I too looked happy. I felt excited. After all, it had taken long enough.

At the same time, however, I was entertaining surprising misgivings. It struck me that there was no separating this clean, wholesome love affair with powerful equipment from the whole continuum of destruction. Even I could understand the narcotic effect. The clean, wholesome love affair extended to every complex and beautiful new fighter plane or nuclear weapon that the federal government commissioned to be built; to every flesh-rending gadget that reached the stage of prototype. All, sooner or later, would be used.

Entwhistle the Younger, a handsome, sensitive-looking fellow, more resembled a symphony orchestra conductor than the artist with heavy equipment. He looked small and sat very straight on the high seat of the yellow, prehistoric monster. It whirled and charged, and whirled again, while he twirled the controls as deftly as if the wheel were a baton.

Next day the son's son, a child of eight, came to observe the digging of the pond and other aspects of the project. Country boys are lucky in that they often get to watch their farmers at work. Three generations of heavy-equipment Entwhistles, putting their minds to it, had the swamp battered and rolled into shape in three days. They had drained and landscaped the upper part of the place, finished a road and circular drive, and dug a pond ten feet deep with a neat little island. They had done a fantastic job.

Mr. Entwhistle's bill, when it arrived, seemed eminently reasonable. I found it hard to understand how, with rates like that, a man could afford to drive a Lincoln Continental.

At first, naturally, I worried that the drainage might prove too successful, causing the whole beautiful swamp to filter away through the ditch. But the next few weeks indicated that he had figured it just right. The warmer side of the meadow and the middle of it dried out enough for growing any kind of fruit trees or crops. The soil would still have to be built up, and the hardpan bored through in places. I could not help but recall Marcie Oleander's confident words about their new farm near the Mendocino coast: once a person had "worked some organic material into it"—with a crowbar—anything could be grown. But still, at the lower end where I planned to start a woodlot and cover for wildlife, and where the pond's outflow ran, more than enough marsh remained.

The Israeli, I had heard, used eucalyptus trees to dry up their marshes. I talked to a Japanese nurseryman and asked him what kind of eucalypti would grow in a swamp.

He opened the Sunset *Western Garden Book* and ran his finger down a list of about fifty kinds of eucalyptus trees.

"Looks like the silver dollar eucalyptus would be a nice one for you to grow," he said. "I can get you a hundred seedlings."

The trees arrived and I planted them. Then they began turning brown. Quite a few died. I dug out my own copy of the Sunset *Western Garden Book* and discovered that the silver dollar eucalyptus "does not thrive in wet soil." Moral: Never trust a Japanese nurseryman who uses the Sunset *Western Garden Book*.

But I found other trees that grew very quickly and did make excellent blotters. Monterey pines, coast redwoods, and white alder, among others. And even some of the eucalypti survived. I also planted a hundred Western redbud, whose cerise blossoms should be beautiful in March.

What the Entwhistle project signified was that the value of the property had considerably appreciated overnight, a smart, unladylike bit of practice as I realized almost guiltily. This emotion was quickly dispelled, however, by the county assessor's lightning reappraisal. One of the disadvantages of living in a sparsely populated area is that a newcomer receives perhaps more than a fair share of personalized attention from such officials as assessors and building inspectors. Even Alice's doghouse did not go unnoticed.

The two appraisers arrived unannounced at Meanwhile one day in a pickup truck, with a pair of rifles racked in the cab. The suggestion of redneckery always causes my hackles to rise. I could not help thinking of Merry England in the reign of Henry VII and the activities of the tax collector known as Morton the Fork (because he carried one) and from whom the expression, "fork it over."

But my two visitors were so courteous and blandishing that I quickly changed my views.

"How did you happen to get the best land along this road?" one of them asked, deadpan, thus setting the stage for another Great Moment in Modern Farming—the arrival of the tax statement. The tax value of our five acres of grazing land had appreciated about 300 percent.

# Chapter VI

# Underground of Over-Achievers

To give nature her due, the Entwhistles had not been entirely responsible for the rapid draining of the field. One reason the water had been siphoned off toward the ditch was that the terrain was honeycombed with the tunnels of the myriad and ancient settlement of pocket gophers.

They watched everything that went on from the moment I arrived. When the farmhouse was rising, there were always one or two sinister little brown heads peering over fresh mounds of earth nearby. The tractor work was observed and no doubt bitingly criticized down in the tribal council chambers. Gopher drainage proved that these rodents also provided another important service, that of aerating the soil. So the main problem would be to persuade them to meet me halfway on a division of the spoils—when there were some spoils—as I thought in arrant innocence.

The farmhouse construction was partly the work of professional carpenters and partly that of a few volunteers who chose not to see me turned black by the first heavy frosts. Notable and loyal among the latter was my old friend the Countess Lillian, who prepared drawings for the Meanwhile farmhouse.

I had first met this dilettante, Renaissance woman, and world traveller on the TransSiberian Railway some years earlier, at which time she was scribbling away at her provocative memoir, *Across Lapland With Skiis and Reindeer*, which was published with a lengthy introduction by a physician who knew a great deal about bilharziasis and seemed determined to get it off his chest.

"Did *you* have bilharziasis?" I asked.

"No," said the Countess, "But I once came down with a disease that afflicts only the polo ponies of Poona, and this proved quite exciting to British medical circles."

The Countess turned out to be an inveterate pilgrim, the sort one keeps running into in the far places—one year operating a smallpox vaccination clinic in the Chota Nagpur region of the Lower Mahanadi Basin, while the next spring might find her on her uppers and selling tiny bottles of Zem-Zem water to Moslems as they debarked from the freighters at Jiddah on their way to the Holy City. She always had a suitcase close by her and this, as I learned, contained old sepia-tinted photographs of the First Everest Expedition. Sure enough there was the Countess in knickers and lace-up boots and layers of duffle coats, carrying her heavy, ancient Speed-Graphic and tripod. As official photographer she had brazened out heaven-knows what ghastly blizzards. Then she had toured the world, lecturing, until her photographs had had to be specially bound in aluminium to reduce wear-and-tear.

At our first meeting I was impressed by the fact that the Countess's hat and frock (as she called it) were scissored from the same bolt of material. Indeed, at this writing they still are, although it is getting harder and harder to find just that shade (fête puce) and grade of bombazine.

On the TransSiberian that year she had been enroute to Australia. Some scheme for popularizing the old Saki dish known as filboid studge. Her family estates in Buckinghamshire and Scotland were going down the drain with death duties, and she had set out to rescue them by revolutionizing the eating habits of the outback country. ("After all, they've subsisted on mutton and sand since the first convict ships came over.") It was a characteristically noble undertaking. She hoped to distribute frozen filboid studge from a Good Humor truck, grabbing the money and running before the stuff defrosted. As she later wrote to me in America, however, the dish was a smash hit. Aussies cried for frozen filboid studge. The sheep posts were gourmandized overnight and for a time one heard of nothing else from Alice Springs to Perth.

Having achieved her objective of restoring the family holdings, however, the Countess found herself unable to settle down.

For a time she became a familiar passenger on the boat from Dover to the tiny Iberian island of Ibiza. There she and a London crony known to Interpol as Moll kept the English community supplied with tins of Crump's Solid Scottish Tea Biscuits, Genuine Bangers, and other nostalgia-inducing contraband. But it was not really daring enough a venture to engage her interest for long. Also, the Spanish Customs officers had begun making threatening noises about picking up one or another of her passports. Thus, when I wrote, explaining the plans for Meanwhile, she was more than ready to chuck everything ("before my cover blows") and move to California to act as a consultant.

No sooner had she settled in nearby San Francisco, and gotten herself appointed Recording Secretary of the Browning Society and a board member of the SPCA, than she drove down to the farm and proclaimed herself ready to draw up house plans.

"But-but," I said.

"Perhaps you had not realized I am an architect?"

"The truth is—."

"Not only am I an architect but I have come prepared to do all your electrical wiring." And in one of her grander non sequiturs she added, "Don't forget, after *all*, that I speak four Tibetan dialects fluently and am the second European woman in history to have been honored with the rank of a Lama."

It was true enough, of course, so I told her to go ahead and rough out a floor plan.

She began to sketch with swift, bold strokes on a scrap of Bronco tissue from the British East India Lines.

I retired tactfully to the kitchen of the construction trailer in which I was then living, fixed her a cup of tea with a tea bag, and brought it back to the coffee table. Without even troubling to look up, she produced one of the spare tea bags that she always carried in what can only be described as a recess of her garment, and plopped it into the cup. "Sorry, luv," she said, "but you Yanks always make your tea too weak."

She had learned architecture, she explained, from her father, Lord Dullchizzle, whose hobby had been carving caskets in the fine Victorian tradition, all woodbine and Virginia creeper. A casketmaker was a joiner, a carpenter, and an architect. Men of this fiber had designed the stately homes of England. Who was I to argue?

"Where did you learn the electrician's craft?" I asked.

"Oh, just a speed-reading course I took in the Children's Room of an American public library," she said. "*The Adventures of Freddy Kilowatt*. You can learn anything from reading the children's how-to books, you know, because they are required to be accurate and extremely simple. It has been a most profitable discovery for me."

She asked whether I wished my Virginia creeper and woodbine carved into the box itself or whether I should prefer to grow it. I answered quickly that I favored the living look.

She began to hum the opening bars of "God Save the Queen," and then said, "My dear, don't trouble your head about another detail."

The drawing for the small farmhouse depicted a bit of a box, as might have been expected—but nothing, as I reasoned, that a monster grape arbor, plenty of fast-growing trees, and a lot of morning glories, sweet potato runners, trumpet vines, ivy, wistaria, scarlet runner beans, honeysuckle, passion flower, clematis, and silver lace vine could not take care of.

The main requirements had been that the house be comfortable, open-seeming, inexpensive, and that it require minimum housework. And in meeting these needs, the Countess succeeded perfectly.

The exterior would be of heavy redwood panels which I felt guilty about, knowing I probably should have used plastic. Each window would disclose a sweep of meadow, swamp, or hillside, and she had thought to provide views of the sky as well via a sleeping loft with large clerestory windows at either end. She added touches that the engineer-designers responsible for most low-cost housing seldom think to include, such as a pantry, and an outside door to the bathroom so the farm dog could come in muddy and clean up before tracking through the living room. The roof would be of rolled-roofing, to save money and trees; and done with the idea of laying on a solar-heating system when available.

Soon the canyon was ringing with hammer blows, and echoing with the whine of power saws, and short, sincere bursts of profanity.

A contractor and his carpenter built the shell of the house. I was astounded by the complexity of erecting even a small dwelling by hand. One forgot, in our age of the modular, prefabricated, sprayed, poured, and sometimes even pumped-up housing, the number of details that must be kept track of in old-fashioned construction; how much had to be known about the stress values of materials if everything were to come out right in the end; about costs and purchasing, and the dovetailing of schedules for the plumbers and other workmen. It helped me to appreciate what the pioneers with their primitive tools had gone through.

One day when the shell was almost finished, Bill Sharpe, the carpenter, came to me and, referring to the Countess, said, "What does she want me to do with that?"

He was pointing to a plank that appeared to be holding up the entire upper front wall of the house.

"That plank don't belong there," he said. "She forgot to put in a beam in the drawings to hold up this front wall. When I take it out, that wall's gonna come down."

He did not look wholly dismayed by the prospect.

I told him to put a beam there and he said, "Well, I thought I'd better make sure before I ripped out that plank."

Events of this sort were not unusual. I had become rather sensitive to workmen who would approach me when my hands were a mass of compacted glue, sawdust, nails, and paint, asking, "What would you want me to do with this?" or "What did you figger to do with that?" This and that always posed problems so awkward that my imagination would be taxed to the outer limits, thus, as I felt sure, gratifying the far more knowledgeable questioner. It was one of the prices one paid for not just turning everything over to the male professional.

After the contractor and his man declared their part of the bargain finished, the volunteer team (if team is not too strong an expression) took over. Applying our skills to the interior walls, we were soon able to create a handcrafted, or dulled-adze, effect. I had never used power tools or built anything before and the friends were almost equally adept. In the crushing heat of Indian Summer, we learned to cut Douglas fir panels more or less straight. We learned to gouge a multitude of wavery little holes into them with an electric jigsaw, to fit around light switches.

The Countess philosophized, "Some things are harder than they look and some things are easier than they look. Putting up wall panels is harder than it looks. Putting knobs in doors is easier than it looks."

I, who was finding nothing easier than it looked, restrained a sharp comment.

A dead giveaway to the current state of the homebuilders' art is the number of materials and methods for concealing terrible workmanship that can be bought by weekend carpenters. We availed ourselves of them all—Scotch tape, fabric tape, wood-patterned tape, wood putty, thick construction glue mixed with sawdust, and a whole line of products with names like Dabble, Boggle-Pruf, Spackle, Dribble, and Dollop. Some worked, some flopped miserably. The last step was smearing over the corrections with two or three coats of paint or stain. All the tapes, by the way, unless they had been painted over, dropped straight off the walls when the temperature soared to 110°—but, for that matter, so did I.

We weekend carpenters were clearly the bane of the building supply clerks. When our kind swarmed into the stores on a Saturday morning one could see them exchanging glances of thinly veiled amusement which, as the day wore on, would turn to sarcasm and sullenness. The discrimination against women carpenters was, of course, flagrant.

A male clerk to whom I was reading off some perfectly straightforward building needs interrupted one morning with, "You'll never believe this! There was a woman in the store last Saturday and she asks for this board, see, exactly thi-i-is long, and thi-i-is wide, and thi-i-is thick; and she was measuring with this rubberband on her wrist—."

As the poor innocent was concluding his miserable anecdote, the

Countess Lillian swept down upon him in her parish hat and the matching floral organza with the mauve crepe de Chine sash. Whipping out her notes, she said, "I should like 2,500 board feet of Merch-grade five-eighths inch redwood plywood, eight inches on center, with no knotholes, mind you."

"M-merch grade?"

"Certainly Merch grade."

"George, do we have M-merch grade?"

Not only did one get prompt floor service when shopping with the Countess, but materials tended to arrive on time, in the proper amounts, in approximately the right dimensions, and of reasonably good quality.

Since I was working alone much of the time, I soon began to learn a great deal about my own stress values. The period of intense heat continued. My fingers were always glued together in paws, causing the thousands of tiny finishing nails I was driving into wall panels to cluster to them as if to magnets. The result was erratic hammering, frequently smashed thumbs, and rage. In my unrelieved gumminess, I developed the melancholy habit of knocking boxes of the tiny nails off the top of my stepladder into the deepening carpet of sawdust and debris. Once I lost a contact lens in this jungle floor. Shrugging, I went on blindly pounding.

The plumbers, Al and George, always arrived at lunch time, a period that I in my haste no longer recognized. They would drive up, get their toolboxes and materials out of their truck, and then Al would say, "It's about that time, ain't it, George?"

"Yep, it's just about that time," George would affirm. "I'll go and get the lunchboxes."

They would settle down in some cool, half-finished room for a leisurely repast. As I worked on through the heat of the day I would overhear the murmur of their conversation, which ranged predictably across such topics as pickup trucks, deer hunting, baseball, black widow spiders, and rattlesnakes.

I would hear Al say, "Boy, when I go under the house after lunch I want that trap door left open, man. Where I was workin' yesterday those black widow spiders were big enough to eat a man."

George would rejoin, "Boy, when I see that rattler, man, I'm comin' right up through the floorboards!"

In torrid weather I stapled fiberglass insulation material between all the wall panels because Sylvia Porter had written that you could save a year's heating bill that way—a message I am still trying to decode since, if you had saved it, you wouldn't have it, so how did she know what it was you were saving? Anyhow, I hoped it was true, whatever, because even with wearing gloves I failed to keep the tiny glass fibers from penetrating my hands, and the red bandanna over my mouth and nose did not prevent them from filtering down my neck, and it was a long, hot job.

On a day when I had decided that for once I had taken on a project that was simply beyond me, physically, mentally, and emotionally, and that it would never be done, a group of successful drop outs from Berkeley arrived, dressed for painting. We uncorked some of Almadén's finest, and soon the redwood stain was spattering around the living room. At last the panels were finished. Another crisis faded into history.

I alone laid tiles in the kitchen and bathroom on a sweltering day. Alice carried my tin glue-spreader out into the swamp and buried it among the gila monsters. Having begun the job, I could not stop, so I smeared away with a wilted piece of cardboard. Soon I found myself literally glued by hands, knees, and bare feet to the floor. Flies were drinking sweat from my face and I could not brush them off. Sweat ran down my nose into the mess on the floor and I was helpless to wipe it. That, I believe, was the absolute low point of the building project.

For possibly the hundredth time I wondered what I thought I was doing out there all alone in that dismal Eeyore swamp. (One mentions these more trying episodes, not to win an outpouring of pity, but merely in the hope of imparting wisdom to other innocents who may be launching Meanwhiles. Line up all the good help you can get, in advance. Promise them anything. And get yourself deeply involved as soon as possible, so that it will be impossible to turn back.)

As the result of more advice from Sylvia, I left the other floors temporary, painting the heavy plywood underflooring with a deck paint. Some Persian and Indian rugs looked fine on this. A black Franklin fireplace was set up on a pad of adobe tiles in the center of the tiny living room. A ceiling-high bookshelf was added along one wall. With the high, open-beam ceiling, the total effect was snug, pleasant, and personal.

Then the Countess bore down on the swamp carrying *The beidermeier Twins Build a 2,000-Gallon Septic Tank*. On that day, fortuitously, the Entwhistles returned to take care of the underground sanitary installations. Another visit from Al and George and the toilet and water pipes were hooked up. Thanks to the indomitability of the human spirit, as I remarked to Alice, everything had worked out just fine.

Once a woman has done something even remotely in the building line, she finds herself in touch with quite an underground network of female over-achievers. The Western U.S., in any event, seems to seethe with them. Plumbing, carpentering, auto repair, every aspect of living, have rapidly become a part of the free woman's stock in trade—and I may say that there is nothing like doing for one's self to build confidence.

Through the network I learned that the pioneers in the reuse of old auto glass and metal for house construction (aside from the shacks of the poor, in which they have long been utilized) were not the Drop City dropouts of the sixties who employed them in building geodesic domes. In the 1940's, "two well-bred ladies from Boston" who settled in Los Gatos, California to build an adobe *casa*, had often raided auto junkyards for materials. I met one of them, Maude Meagher, after reading with admiration her book on building Casa Tierra, by far the most handsome adobe construction I have seen. She and Carolyn Smiley planned it as the headquarters for World Youth, Inc., an international student group they had founded in Boston. Today it serves as the home-cum-laboratory of a scientist.

Shortly after Meagher and Smiley began their six-year undertaking, World War II claimed all ablebodied workmen and most of the standard building materials. The two women found themselves making thousands of adobe blocks, which weighed from fifty to seventy-five pounds each, and hoisting them up into place in the walls of the rambling, Spanish-style house, which they had also designed. Both were completely inexperienced at such work. Mixing clay in an old cement mixer, building everything from the earth up, literally out of earth, they managed to construct walls. Just as it was time for roofing, they were offered some railroad carloads of Spanish tiles that a contractor had no other market for. But no glass was available for the multitude of windows. Meagher salvaged it from old cars and laborously learned to cut panes to fit. In all the main rooms they built and tiled enormous fireplaces. For the fireplace in the meeting hall, they ingeniously employed as tiles the individual photographic copper plates of children who were members of World Youth, Inc. In their small book describing the project, the frontispiece bears the quotation: "Whatsoever thy hand findeth to do, do it with thy might.' "

There were other triumphs of shelter-building by an earlier generation of women which I was told of, and some of them I visited.

Barbara Nelson took me to see a family vacation cabin that her mother and three of her aunts had built, like a Tibetan monastery, high on the side of a cliff overlooking Puget Sound.

"All four sisters were in their sixties and seventies," Barbara said. "They started the cabin casually one weekend, using a copy of the

*Ladies Home Journal* as a square, and a rock on a string as a plumbline. None of them had any building experience. My mother was the vice president of a bank and Aunt Anne was a schoolteacher; some were married, some not. But they had a good sense of spatial logic among them, and all the confidence in the world.''

Aunt Anne took care of dynamiting the building space on the cliff ledge "because she had access to a book on the subject." In the future, whenever she was asked for identification, she was pleased to show her blasting permit rather than more mundane documents. She sounded a lot like the Countess.

The sisters bought an old Seattle streetcar and bashed it up, salvaging windows and metal in the frames for the four-room cabin. Before the roof was on, they lowered down the cliffside a cast-iron stove, a refrigerator, and other heavies, using a block-and-tackle—which they later employed in digging the well.

When the husband of one of the intrepid quartet finally became convinced that the vacation place was going to be finished, he offered to come the following weekend with another man and the necessary scaffolding, hoists, and other equipment for putting on the roof sheathing and shingles. But the foursome, true to the impatient spirit of Ruth Stout vs. Plowman, saw no reason to sit around cooling their heels. The day the men arrived with the truckload of heavy equipment, they were slightly annoyed to see four little white heads peering up through the last open holes in the roof. The sisters, with the inventiveness that sometimes blesses those who have been spared doctrinaire instruction, had simply stood on ladders inside the house and nailed the sheathing and shingles on while leaning out through the open roof.

They did all the wiring and plumbing, and dug a septic tank. A connecting bunkhouse with several beds was added, and a deck with an outdoor firepit. And having finished the whole job, they thoughtfully named the cabin Dead Reckoning—in honor of the fact that it had been paid for by a timely bequest.

Aunt Anne then went on to greater achievement, for it turned out that she had a flair for witching wells. She got vibrations on an adjoining ledge, which belonged to a vacationing banker and his wife, and glibly persuaded them to let her drill for water. Back to the old block-and-tackle. She struck good water at forty-five feet.

The Meanwhile farmhouse, compared to projects like these, had been no more than an Indian Summer's interlude, an exercise with finishing nails. But it too was solid. God willing and the swamp didn't rise, it would stand for quite a few years.

Chapter VII

# Main Street
# & the Pickup Mystique

On Main Street, U.S.A. in the early seventies, a newcomer had good reason to expect the local populace to be wary. It came as a pleasant surprise to discover that old-fashioned warmth and trust still survived. It was I with my uptight city ways and compulsiveness who sometimes had to make a point of slowing my reactions. Also I had to relearn that generosity of spirit need not be suspect.

Among the many surprises that awaited me was the discovery of craftsmen who still struggled to achieve perfection in their work, and of crafts I had not even known existed.

During the building of the farmhouse, I went to see Mr. Thomas, who made prehung doors. He made them for two large counties and told me that he wanted nothing more in the world than to get out of the prehung-door business, but that the two counties had no successor to leap into the breach. In the eyes of youth, it seemed that the glamour had gone out of a career in prehung doors; or more likely, few people ever wondered how it was that some doors fitted perfectly while others stuck or sagged. I was among them. My lifelong attitude to doors, I fear, had been that if one failed to open, you could usually find another to pound on, or you could call a fireman with an ax. This brittle attitude I quickly came to regret as Mr. Thomas opened my eyes.

He said that, short of being allowed to retire, he would have been satisfied just to find an assistant, but that those he tried never measured up to anything like his standards of perfection.

"When I say I want a door hung to one-sixteenth of an inch," he said, pushing up his plexiglas goggles and flicking a small mountain of sawdust off his nose, "I don't want it going out hung to one-eighth of an inch and having it come back tomorrow."

·"I should think not," I commiserated. Then I proceeded to give him the dimensions for a pair of doors I needed. They proved to miss, by a couple of inches, fitting the holes they were supposed to fill. They had to be brought back and exchanged for a completely new set, greatly to Mr. Thomas's professional distress.

His was certainly not the sort of hue-and-cry a woman often heard from a city craftsman.

He had a sign over his front office that said, "IF YOU DON'T SEE ME HERE WELL LOOK OUT BACK," with two eyeballs in the O's. And that was where he was usually to be found, working at his circular saws in a more or less constant shower of extremely fine wood shavings. It was where I found him that first day, not only to give him the wrong measurements for the new doors but to have him prehang and put locks on a pair of old French doors. He told me he was just on his way to the chiropractor's.

Remembering my problems with Entwhistle I said, "I expect these'll be ready about next Tuesday?" I made it sound casual.

He laughed, forgetting his aches and throwing back his head.

"You won't be needing these doors for three weeks," he said.

I stared at him, thinking this a fine piece of impudence.

"How do you know that?"

"Because I know who is building your house and how far along it is, and how long it takes workmen in these parts to do certain things. And you aren't going to need any prehung doors for right on three weeks." He grinned.

Good grief, I thought. No wonder he was overworked.

Bidding him goodbye, I walked over to call on the local agent for my car insurance, relishing the idea of the lower rural rate.

I went through the door, saying, "Hello—I'm . . ."

Mr. Elwood Thrigby bobbed up from an open filing cabinet.

"Oh, *you* must be Mrs. Cheney. I was just going to telephone you. Have your file right here . . . someplace . . . My secretary's out just now, but it's—right around here—."

I was beginning to feel as if my life was an open book kept in a glass house.

Mr. Thrigby said, "Well! Are you finding your new surroundings inspirational for your work?"

"Do you mean hammering up the wall panels, or my definitive hysto-biography of Norman Mailer? Because if you mean the latter, negative. I've had to put it off until I can figure out how to typewrite on the head of a pin."

He said something about understanding because he was "a creative person" himself.

That was all it took—the least hint of unexpected sympathy after those two or three exhausting months of struggling to get the swamp licked into shape. Suddenly all my frustrations and trials erupted. I began gibbering.

"I haven't really—what with. And waiting for the prehung. Getting the doors dredged. It takes. It all. Somehow. Everything takes so much *longer*. From scratch. But I suppose. You. Actually. It's funny. Why am I *crying?*"

"There, there," Mr. Thrigby said.

"It's all a muddle, Mr. Higby."

"Thrigby. Elwood. And I *do* understand. Because *I* keep thinking of so many plots for plays all the time that it drives me crazy!"

"No!"

"Yes!"

"But that's marvelous."

"Do you *really* think so?"

"I *certainly* do."

"Sometimes," said Mr. Thrigby, "I get ideas for half a dozen plots a day. I even get them before breakfast. Tell me—what do you think I should do?"

"First," I said, "write them all down."

"Yes, yes. And then?"

"I'll have to think that over very carefully," I said. "But in the meantime you could give me a refund on my car insurance premium because of the rural rate differential."

"Oh, they'll have to figure that in the district office," he said. "But I'll telephone you just the *minute* they let me know."

And Elwood Thrigby was as good as his word—or rather, better, for he telephoned me the very next day. He wanted to tell me about a new plot.

In small towns many important business deals are still transacted by oral agreement and sealed with a handshake. This is not necessarily an outgrowth of excessive honesty. Partly it is because we trust each other more than we trust a lawyer.

Another reason is that, more often than you might suspect, at least one of the parties sealing the bargain is virtually unschooled and therefore understandably prefers the handshake to a piece of paper. This has nothing, however, to do with an abundance of native wit.

Tim, for example, who is almost unable to read English, is a quick study at figuring out any sort of technical problem and is reputed to play a sly game of chess. Angus, who is now a contractor, followed the harvests through Steinbeck country with his Okie parents at an age when most children went to school. He has developed his memory to compensate for a limited ability to read and write, habitually going around with his skull crammed full of accurate dimensions. Someone can always be hired to take care of the bookkeeping.

The quality of trustingness is one that the new and righteous revolutionaries such as Tom and Marcie Oleander have in common with the shorthaired, cleanshaven townspeople. Alas, it is not reciprocal. Merchants tend to trust just about anybody but the "longhairs," and it seems inconceivable to most of them that shaggy mops can conceal all kinds of minds. It is all right to have your hair neatly trimmed to where it just brushes the top of your collar, but anything beyond that is likely to be seen as a symbolic rebuke of clean necks and patriotism.

A reasonably square-looking stranger can cash a check for goods purchased almost anywhere in my area without first being run through an FBI screening, suspicion tending to be aroused only if one makes the mistake of *asking* if a check will be acceptable. That marks you for a stranger, Stranger—.

Our region is not without crime, but so little of it occurs that you almost always know exactly where to find the local law. Nine times out of ten if, say, some known teenager spraypaints your mailbox, you can find a posse down at the Sno-Joy on Main Street. They will be leaning against their patrol cars with the radios crackling, drinking coffee out of plastic-foam cups and arguing high school football. Citizens try not to report any high crimes or treason during the normal coffee breaks, and certainly not if they coincide with a game.

In the canyon I always feel especially secure just before the opening of deer season. At that time of year I sometimes glimpse the constable's car patrolling very slowly up and down the road at dusk, flashing a beam up into the brush. I sleep more soundly for it.

If you happen to be driving on the county road and one of our official cars streaks past with its amber lights flashing, the chances are the officer is running a little late for the Sno-Joy. In that event: 1) two pickup trucks probably have crashed head-on out on the straightaway of the two-lane highway as the result of one of them trying to pass a third pickup truck (cowboys and ranchers are notorious for overestimating the pickup of their pickups); or 2) a teenage boy, driving his father's pickup, has lost control on a long curve and wound up dead in a drainage ditch. An astounding study disclosed that a fourth of all the accidents in the county were of this one-car variety, involving youth.

Rural people, rural space, and pickup trucks make for fascinating interactions.

For example, like our city kin, we may occasionally harbor personal grudges. A convincing case exists *against* high-density housing, however, in the ample evidence that bad blood is filtered out through the buffer zones of rural space. Two enemies, if they choose, find it easy to avoid each other. Although they may have to meet once a year at the sheep-judging or the pruning demonstration, this can be coped with short of developing high blood pressure.

But even people who haven't spoken for years (for so long they have forgotten why), go on waving at each other's pickup trucks. Not to wave at a neighbor's pickup truck is condemned as inhumanly cold and malicious. After all, it wasn't the truck's fault that it happened to be bought by that sneaking so-and-so. There is the further consideration that trucks change hands and people acquire new ones. The pickup you failed to wave at yesterday may have been driven by one of the supervisors or the bureaucrat in charge of federal subsidies. So we play it safe. Pickup truck etiquette also extends to passenger cars, and for the same reasons.

The normal serenity of country living imposes another responsibility on the pickup owner, which is that a good neighbor tries to carry something reasonably provocative in the back of the truck.

When two pickups meet on a narrow curve, both going hell-for-leather, the trick is to be able to wave and sneak a quick look at whatever load the other person is carrying, as well as making a literal value judgment about it and a shrewd guess as to where he got it without (unless you are a teenager) slewing off the road.

Since I have no pickup truck and am usually just poking along, gaping at old barns or new lambs, I cannot really count myself a part of this sport. In fact I count myself so far out of it that, whenever I hear two pickups hurtling toward me, I simply shudder to a stop in my scarred compact wherever I happen to be, as a public service. By causing my neighbors to slam to a stop, I am better able to appraise what they are carrying and also—as I have little doubt—I help to save lives.

One reason we carry interesting things in our vehicles is that we all have to take our garbage to the county dump. Long before city people discovered recycling, the rural dump was a lively trading center. A farmer knows that, sooner or later, she will have a use for any found object. Not its original use, probably; but there are a thousand farm needs for which one must improvise (and personally, I hope that U.S. manufacturers never discover what they are).

Some garbage-dump operators make a fair penny salvaging and selling. The local antique dealers' trucks may be seen parked nearby. Such commercialism takes a lot of the pleasure out of an honorable country pastime; but even if you have to pay a dollar for a broken oak rocking chair or a stuffed screech owl, it is worth it just to diversify the load in one's pickup.

It can be *important* to know who is going where with what. Say Ed Barnes sails past in his pickup with four sacks of Eggena and two lambs that he must have picked up at the auction sale. I immediately wonder if the lambs have Gatley's Blotch like the last pair, and Ed is planning to use them for an income tax loss. But let us say, for the sake of our dramatic narrative, that his pickup contains three monstrously long steel pipes, with a flash of red cloth at the end. We are an irrigating community. Anything that concerns the transportation of water interests us intensely. Ed is going home with *used* irrigation pipes. There is no way under the blazing sun that he could have gotten them at this time of year unless some farmer is going broke—or has made so much money that he is selling everything including his rabbits and moving to Las Vegas to take up with Bunnies; and in either event, bankruptcy or nirvana, it is likely he will also be selling his wooden trays, his Black-welder Gopher Machine, his harrow, disk, seed-trencher, tractor—the works. So it would be smart to hurry on home and read the Classifieds.

A traditional after-supper pastime of rural menfolk is to stand around outdoors when the weather is decent, comparing and coveting each other's pickups. They smoke a lot of cigarettes. A pickup, in addition to its power/noise gratification, is generally felt to be more wholesome or cleaner than other forms of emotional outlet. The camper-trailer is *okay*, provided it has motorcycles and trail bikes for the whole family racked on top; and almost any kind of new, expensive wheels will guarantee you an after-supper circle. But, say what you wish, year in and year out the pickup reigns unchallenged out here where the deer and the yuccalope play.

"If it's so wholesome," the goodwife may ask, "what were you all doing out there in the dark?"

"Smoking, my dear, so as not to stink up our home."

While stalling for twilight, they engage in mannerisms that would not deceive an undergraduate sociologist. There is much nervous plucking at the fleece on chaps. Stetson brims are rolled and rolled again. So many sulphur matches flare that you could swear it was the Fourth of July.

They crack their knuckles. They snap match stems between lean, brown fingers. As night draws in, one of the group drifts forward and sneaks a fast kick at a radial-belted-super-nylon-glass-steel-ninety-ply tire. He fades back. If it comes to a bust later, they'll have a hard time fingering *him*.

An anonymous, buttersoft voice drawls unrestrained endorsement. "Yep!"

Suddenly an oldtimer who rode shotgun in the big pickup roundup of '06 leans forward. In a gesture of bravado that would be tolerated only in him, he runs a leathery palm along the tactile fender, feeling the prime-coated, oven-baked, metallic high-gloss enamel. A shiver trembles on the patriarch's lips. He shouts, "Ya-*hoo!*"

Again the matches flare like freaked-out fireflies. The air, like the mood of the men, has turned sultry. Summer lightning zaps suddenly at the sleek, 49.3-cubic-foot chassis. Beneath that hood, as everyone present is acutely aware, lurks enough coiled power to get you from 0-50 in 8.0 seconds at 34 mpg, give or take a gallon or so.

A rawhide voice, previously uncommitted, utters the judgment that legions of Western womanhood presumably long to hear, if in somewhat adapted form: "Best motor GM ever made!"

The dramatic tension is shattered by a call from the kitchen door. "Melvin! You get in here and empty the garbage."

Recently I drove to the gas station at San Solace to find a very impressive new pickup parked over in the viewing circle. It was attached to a huge trailer, a cross between a horse trailer and a livestock carrier, if indeed it were not a travelling zoo. The sides were covered with heavy steel mesh. Surrounding it, their flushed faces moist beneath straw hats and Stetsons, the Marlboro cigarette packs wilting in their pockets, crowded a ring of men and boys.

When Joe the station manager reluctantly left the circle to come fill my gas tank, I asked him what they had in there.

He grinned in embarrassment, an X of wire grill impressed in the tip of his blunt, honest nose. I began to wish I had not asked.

Blushing, he blurted, "Oh, nothing. This guy Ed Barnes. He's just bought a new rig, is all."

Chapter VIII

# Making the Adjustment

To be sure there were a few hardships to adjust to in the country, times when I missed the challenge of random thuggery on city streets, the throb of commerce and culture. Most of all perhaps, I missed the intellectual ferment of Berkeley's Telegraph Avenue at the nostalgic hour of 10 o'clock when my favorite paperback bookstore proprietor yelled, "Okay, everybody! Mark your places!"

When our country Welcome Wagon Lady dropped around, she sounded me out on organizations to which I belonged. I have grown cagy with the years, skillful at obfuscation. "Just the Audubon Society, the League for Better Government, and the Orphan Annie Union for Teenage Sexual Promiscuity." After a moment's silence she said thoughtfully, "I guess the closest thing to that would be the Great Books Club over to Fat City."

Usually the first question city people ask when you announce a move to the land is, "But won't you be lonely?" The answer is, "Only for you, baby." The second is, "Whom will you talk to?" (Again you give expected answer. "Iks.") And the third question: "How will you keep in touch with the world's affairs?" ("Ukk!") These answers are based on the age-old conviction of urban folk that civilization ends abruptly just across the Great Divide from their own job-centered universe. From there on the world is flat, politics nonexistent, and news is communicated by means of word-pictures drawn in the earth with sharpened sticks.

They forget, these beleaguered souls in the polluted hearts of cities, or perhaps have never heard of, that lively and informative medium, the sticks.

And we also have television in the country—two whole channels.

As a rule I find it extremely difficult to make a choice between the one that favors us with virtually nonstop regional revival meetings generously sponsored by fundamentalist religious groups, and The Other One.

The Other One (or TOO) provides access to a hearty, reliable, intellectually stimulating fare of high school football, college football, pro football, baseball, boxing, handball, hockey, marbles, tennis, golf, swimming, pingpong, mumbletypeg, wrestling, soccer, volleyball, high jumping, pole vaulting, high and low hurdling, sprinting, dashing, javelin throwing, shot putting, basic handstands, and our perennial favorite, Celebrity Bowling.

Then again there are times when we receive the Emergency Testing Signal.

We women, who make up 53 percent of the consumer audience, are grateful, you may be sure, to our programming patriarchy. *Mens sana in corpore sano*, as is writ above the locker-room doors of every gymnasium in the land—roughly translatable as, "The men's sauna is sanitized."

We were very lucky in our area in being spared the Nixon impeachment proceedings because of an extremely controversial Little League series raging at the time. Nor is it unusual for us to find the half-hour of network national news preempted by the County Frog-Jumping jumpoffs. And it's just as well. We homemakers have little time for the tube. What *we* say is, "A used mind is a tired mind and the brain is only a muscle."

Of course, we're not limited to our weekly newspaper and a couple of TV channels. I get my Hoedale Press bulletins regularly. Here's the headline on our lead article in No. 2233: "THE EUPHORIA IS NOT A TRUE SUCCULENT BUT THE PARTHENOGENESIS LIKES PLENTY OF SUN." Right next to a hot tip from Aunt Flopsy for all you ace turfsters out there who may want to get rid of your athlete's foot: boil up a good mess of cudweed or rabbit tobacco until you've got about a gallon of the tea, and drink it right down. Won't have *no* more trouble. Aunt Flopsy swears by it and she's still batting a thousand on the Japanese Indian-Wrestling Circuit. And how about this? Every copy of attached coupon, it says, entitles you to receive one of Maudie Muller's gigantic nocturnal mandibles crossbred with naked sunflower seed by our local Experiment Station. Here's another: "This bulletin supersedes Leaflet 448, Growing Onomatopoeia in its Wild State, 0-292-498," with a sidebar worthy of a main feature: "A good ground cover is dichotomy but if you really want to get into it, just try hybridizing pudenda with any old addenda, then pruning and pinching back severely. What you'll get, unless our Future Farmers are 'way off the mark, is a mixed ingénue. Give it a whack and check it out with Aunt Flopsy or old Maudie Muller."

No, we're never at a loss for information input.

"Won't you be lonely?" "Whom will you talk to?" "How will you keep up with the world's affairs?"

Okay.

Straight answers.

When I made the transition, I managed to turn completely away from electronic devices. I was tuned into the rare and sobering experience of time that *seemed* to lack finity and of silence that had the quality of stretching almost forever. Neighbors that first year were few and far between. Since few of us ever encounter unlimited time and silence, the experience is unforgettable.

Time, pulling back, abstracted all the friendly crutches of habit as well as the tyranny of deadlines. This resulted in almost total freedom, a menacing condition for which nothing in the past had prepared me. I understood how prisoners in solitary felt—even though it was possible for me, by an act of will, to escape and find company.

Small habits such as brushing my teeth, cooking a meal, eating it, walking, provided a sense of security in a structureless new life. These were devices to fend off panic.

If my life was to assume form, I must make it. I would neither be praised for doing so or blamed for not.

Had I, for example, even once allowed myself to start screaming against the encompassing silence—a possibility I entertained—I felt I should not have been able to stop. Occasionally in this strange world without time and the sound of human voices, lacking the hum and whirr and click of routine household appliances, and yet in which I refused to turn on radio or television, I had a vague sensation of falling—not down, but across space—drifting but never landing.

It seemed to me in this period that the artificial measurement, time, and the natural, thick stuff of silence were quite similar qualities or states of mind. They seemed to exist only by their limitations. For example, if either was abundant, it became virtually indescribable—because we know them normally only in terms of scarcity. In driblets they are easily characterized: "I meditated for half an hour and envisioned—." "For five minutes I dreamed of—." "The jackhammers stopped outside the window at half-past-three and—." Images at once arise. A loudly ticking clock illustrates perfectly, making both time and silence finite. Instantly they are comprehensible.

Early impressions of solitude at the farm reassured me that it was in no way to be confused with loneliness, the unfilled need for reaching out and touching. This is not to say that people cannot be lonely in the country, but merely that at Meanwhile, I was always touching and everything I touched breathed life.

Total freedom, which of course had been no more than a passing illusion, lifted its weight from my shoulders the moment I got into the farming swing of things. Another weight descended then, yet not unpleasantly.

I became aware that a row of nagging, almost-new tools leaned against the wall outside the unfinished kitchen door, that seeds and bulbs, seedlings and fertilizers were there; that hard work awaited while the season, in its rude haste, would not. For the first time in my life I became aware of the ominous significance of the dates stamped on little seed packets: "Zone 2, Mar. 1–Apr. 15 . . . ."

As city mouse, I had found whole seasons passing almost unnoticed in my preoccupation with personal affairs and in the general screening out of nature. The discovery always brought a kind of panic. If a lifetime could be passed that way, unnoticed, then which was life and which death? Here, however, the seasons were both friends and bullies and to ignore them was not only impossible: it was to court disaster. One might think to shed a tyranny, but it was never anything but a trade.

This new business with nature had left me momentarily stunned. She, if at all concerned about me, must have been merely baffled. People accuse nature of being many things, including an inefficient wastrel. I found her ruthless and cunning, with scant regard for fools, but also at times surprisingly accommodating. She accommodated beyond what I should have considered the call of duty, that is, of natural necessity. She acted just for sheer delight.

Nabokov has noted this where butterflies are concerned: "When a butterfly has to look like a leaf, not only are all the details of a leaf beautifully rendered but markings mimicking grub-bored holes are generously thrown in. . . . I discovered in nature the nonutilitarian delights that I sought in art. . . ."

Consider the most ordinary example—the habit of deciduous trees of opening their leaves in springtime, shading us through burning summer and, at the first frost, when our heads no longer need a screen, dropping leaves in extravagant swirls for wading through, kicking, rejoicing in. Since trees with needles stay green, it would have been easily arranged for leafy trees to do so—as in fact a few do—in winter. I had seen the deciduous trees go through this exercise all my life. Yet only now, when suddenly across the plowed earth in autumn I could see for vast distances, spotting whole new farmsteads in a world that had not before existed, seeing some perfect jewel of an old red-and-white barn beyond a broad pasture, only now did a great light dawn. It could only have been planned for the personal delight of the lowliest grub with eyes to see.

And the great thing about it—to quote someone or other—was that, insofar as a viewer's contribution might be concerned, "there was no damned merit to it."

The utilitarian aspect, although it existed, seemed peripheral. If I raked the leaves that had given shade in summer and opened up a fresh vista for winter, they would rot on the garden for spring, restoring fertility to the soil. A seemingly simple business and perhaps not as efficient as several solemn men with a computer might have worked it out; but what truly intricate planning underlay it. A day came when a single leaf fell. All was made clear.

So many examples of nature's guile soon become obvious in the country that it is hard to name only one. Consider, for example, her fiendish cleverness in making garlic an asexual vegetable. Had garlic required a mate of the opposite sex, how long would it have survived? Nature's trick for insuring the perpetuation of this herb is almost as sly as the device of the orchid that so resembles a certain pollen-carrying insect that foolish insects try to mate with it. Mother Nature knows that, sexually, it takes all kinds. Only Biolical man, poor ego tripster, suffers from the illusion that his is the One True Way.

Folklore and superstition, as the landperson will quickly tell you, are vitally important. Young farmers may go off to the state university to study entomology, horticulture, plant genetics, or agribiz 24-C, but when they get back to the farm, if they have any sense, they flip open the *Farmer's Almanac* and plant by the phases of the moon. Every successful farmer I have met does so.

Personally, I should not dream of letting a Lammas Day pass without spitting on a crooked coin and slipping it into my purse. This wile has protected my apricots from brown blossom rot and my potatoes from galloping seedwarts—at the very least.

Being able to tap just the right homily in a pinch can spell the difference between survival and losing your shirt or, possibly, breaking your neck. I learned to keep a few sayings ready-to-hand in the hidyholes of my head for whatever emergency. When the fireplace chimney smoked, as it often did, I would get out the ladder and an old whisk broom, climb onto the roof, teeter up to the ridgepole and whack the screen at the top of the chimney. This dislodged fragments of charred paper which the rain had glued to it, permitting smoke to escape once more—or rather, it did if I remembered to mumble a few words. What *I* usually muttered for a stuffed-up chimney was, "Blown fuse in the morning, sailors take warning."

Creature of my times that I am, however, I believed in combining magic with the highest technology available. Also, I worried about how I would keep the fireplace from smoking when I became too old and infirm to creep along the ridgepole. So I telephoned the man who installed the Franklin fireplace.

"Would it help if I used a screen down near the flue at the base of the chimney?" I asked.

"Wouldn't do no good," he said. "But there's a chemical you could burn in your fire that would dissolve the soot and ashes up in the chimney."

"Okay, I'll try it."

"I wouldn't if I was you. It'll drip down in a rain and make a big mess all over your chimney."

"Do you have any other good ideas?"

"Well, I'll tell you what I did—I've got a fireplace just like yours in my cabin up in the mountains—and what I did was climb up onto the roof with a stick and knock the screen right out. Didn't have any more trouble."

"That's a great idea! I think I'll do that."

"But I only use my cabin in the winter when there's no fire danger. Smokey the Bear wouldn't like you if you did that, where you live."

"I see."

"Now, here's another idea. You could try climbing up onto the roof when she smokes and whack her with a whisk broom. Of course, you have to keep nimble—."

"Well, thanks a whole lot."

If one's problem is a piece of complex household equipment that has broken down, it can pay to call a repairman and *combine* his services with the best soothsaying at your command.

We find, as we learn more about technicians, that even they rely on remedies motivated more by superstition than logic. Hence I never hesitate to toddle around in their wake, muttering my old-crone's magic. Example: When I called the water-softener repairman because my "unit" was humming (which seemed provocation enough to request a forty-mile round-trip house call under the service warranty), the expert arrived in what appeared to be a fit of genuine pique and whammed the machine hard with his fist, causing it to quiet down for the first time in a week. No sooner had the humming stopped than he could hear me right behind him, mumbling away about icycles and crooked snowballs making Jill a Frigid Girl—which is merely the obverse of that old Southern business about seeing a blacksnake and losing your manhood, as *Foxfire* readers will recognize.

"There!" I exclaimed. "Worked like a charm."

To my astonishment, the technician began waving his arms, shaking his pipe wrenches, rolling his eyes, frothing around the edges of his beautifully capped teeth, and otherwise acting in what might—if I were writing it all up in a social worker's report—be described as a highly disoriented fashion. Thinking first aid was demanded, I dashed into the house for my asafetida bag and a sterile avuncular. Well! by the time I got back outdoors, all I could see was the end of his coattails flapping down the county road. He must have almost seen a blacksnake.

One of our native Californian sayings is, "When you see a tarantula crossing the road, it'll rain in seven days." This bit of folklore happens to be true and, for us farmers, extremely important, because the only time you're likely to see a tarantula at all is at the end of a long, parched summer. We shall be examining tarantulas in a subsequent chapter, not just to find out why they cross the road but why they go right down into holes on the other side of the road—considering that they have just come up out of holes.

A swinging homily wants a bit of mystification and foreboding, as I had learned during an enlightening period when I lived on a farm in England. Charles Bunger, the Kentish gardener who came with our leased premises, and an incredibly spry octogenarian, was a master of this art form. From him I learned that the most potent sayings were motivated by a generally gloomy outlook mired in the anticipation of specific disaster. When Bunger sounded happy, our household trembled.

Yet one of my favorite Bungerisms was the disarming: "Plant yer broadbeans on the first o' May and then go to th' Tenterden Fair."

Accordingly when May Day rolled around, he sowed a row or two of broadbeans. At around ten o'clock he put away his bill 'ook, his baggin' 'ook, and other Kentish appurtenances designed to amaze, mystify, and cow the invading Yankee imperialist. He rolled his bike out of the barn, snapped a clip on his off trouser leg and, with a yank at his ginger forelock and a merry Hi-Ho! he was off. I gathered that the loyal subject would as soon have dreamt of working on the Queen's Birthday. The Fair sounded so good that I decided to go myself. Only thing— when I got around to inquiring directions in the village, they told me that there had been no Tenterden Fair since the terrible marrow epidemic of 1902.

Another Bunger homily for the same occasion (he was truly shameless): "Shed yer coat the First o' May and ye'll give a friend away."

"And what might that mean?" I asked, alert to trickery.

"Ar!" he choked in helpless mirth. "It's only a old saying to mean ye're apt to die o' pneumonia."

Usually he concentrated on the anticipation of misfortune, pure and simple, expressing it in the form of a negative wish. He did not fool me for a second. For example, when the tall fireplace chimneys smoked, he invariably said, "I 'opes yer ain't got a rook's nest in yer chimney." And when he arrived one morning to find my husband and me grimly mopping up the living room floor after an extensive leak from the second-floor bathroom: "I 'opes yer ceiling won't fall down."

Sometimes his warped hopes were openly expressed. For example, during that bitterly cold winter he never tired of declaring, as he held the kitchen door wide open, "What we needs now is a good freeze ter kill th' germs." All the English, of course, believe profoundly in the virtues of a punishing winter. You would think it something laid on by Her Majesty with their highest interests at heart and that God would strike them dead if they forgot to be properly grateful. Sometimes God struck them dead anyhow—from frostbite. Wholesome, germ-free frostbite.

Thanks to mental alertness during this period of my life, I was able much later to settle in at Meanwhile with some lay experience in greasing the farming operation with homily and superstition, the chicken fat as it were of the rural occult.

But oddly enough, although I searched, I found no hex or folklore that applied specifically to earthquake country. Perhaps this was because the state of the predictive art was itself primitive enough to satisfy an average child of nature. Every few weeks or months, I would be rudely shaken from my ruminations by rumbles from deep down beneath the earth's crust. The ranchers claimed that their cows were more reliable earthquake predicters than the university's laboratory equipment. Such animals showed distinct signs of restlessness before the onset of tremblors. But I had decided not to keep a cow after reading what one ex-cattleman (and perhaps ex-con) had written: "The only difference between keeping a cow and being in jail is that, in jail you don't have to milk the cow."

Finally, in desperation, I turned to *Poor Richard's Almanac*. Wasn't much old Ben wasn't into. And thereafter, when the earth began to roll, the boulders to grind against each other and the new, small farm-house to rock, I would mutter, "One Today is worth two Tomorrows." (Knock on wood.) And you don't have to live in earthquake country to appreciate that.

As I settled in that first autumn at Meanwhile, I sometimes fretted over what had seemed the odd reactions of a few friends and acquaintances when I had first begun searching for a farm. What I had expected was, I suppose, a simple if pitying, "Good luck."

One old pal discouraged me from the start by saying, "Even small places in the country are terribly expensive now."

"I know," I said. "But I plan to keep on looking, because there have to be places left that are marginal." I kept using words like *marginal* and *minimal.* "When I find this marginal farm, I plan to live at ab-solutely minimal cost."

She replied, "There's no way to live minimally these days." (Not, "In my opinion there's no way" but just *No Way.*)

After I had found the swamp and had happily divulged the news to old pal, her response was, "How can you bear to live inland where it's hot instead of by the ocean?" I replied that the ocean was only twenty miles away and that its fog would cool the farm each night. She then asked, "What will you do when terrible things happen, as they do on farms?"

"Like what?"

"Well, when I grew up on a farm, we had a pregnant cow that got her foot stuck in a spittoon. We had a frantic time getting it off. If we hadn't managed to, she would have aborted her calf."

The ingenuity of her protest pleased me. I said if it happened to a cow of mine I would probably try to enter her in a special category at the county fair. Best Unpregnant Cow With Spittoon on Foot.

The next alarm was, "But won't you feel terribly lonely?"

This made me furious because I was afraid of just that.

"Certainly not!" I snapped.

As I went on with my plans, other things said or unsaid by this per-son whom I had thought one of my best friends made me suspect that in her head she had even begun to persuade herself of some sort of fic-tionalized angle that might explain my irrationality. Perhaps, forced to leave town in a hurry because of an advanced case of hereditary disease. At last I realized that she, having convinced herself that I was doing a dumb thing, would now be able to continue to live her own life in the old routine, bored as she often had told me she was (in a prestigious, high-paying job), but now feeling that in any event she was better off than some people who had the crazy idea that you really could make a drastic change in midstream.

At a party I was introduced to a man who, on hearing of my plans, simply stared at me the whole evening as if I had popped down by UFO. He seemed to be watching me clinically, and waiting. Another acquaintance asked, "Are you going to become a recluse?" I answered, "Yes."

The sad thing was that I had not departed in a spirit of abandonment of my friends or disparagement of urban life, for I have always loved cities and one or two with passion. I had merely bartered a complex environment for the time and quiet in which to do my work and to live in a way that for me now made sense. At least I hoped that would be the case, although my uncertainty was great. So I fretted until I came on a book that helped me to understand. Psychiatrist Davis S. Viscott writes, in *Feel Free*, about how "people whom you would not have expected to react that way will be threatened by your move." They have reasons why they may be unable to change their lives or even to face certain unpleasant truths. It's all right to complain a little about the system as long as you don't rock the boat.

The boat-rocker is saying, in a sense, that the friend's way of life is not optimal either.

As he adds, all too accurately, "What you will hear is a recitation of the reasons and rationalizations they use to keep their own minds in place. This is a very tiresome business."

Perhaps I should have been surprised, instead, by the staunchness of those who rallied around in every way. Some of them, half my age and hoping to make the break themselves, came to call and stayed to help. I knew I was letting them down, however, for I was compromising more than a little—with Pacific Gas & Electric Company, the telephone, and all the rest of it. I had no time for things like washing clothes the romantic way in a tub heated over a woodstove, with soap handmade from tallow hand-rendered from hand-raised shoat. Much as I respected the young for their return to the basics, fully appreciating the need for each generation to come to grips with physical hardship, my own generation had been through it. Perhaps I would dabble in the arcane herbal remedies, but when the chips were down I probably would call my doctor for an appointment. I would avidly read about pennyroyal and fenugreek and even plant them; but when it came to making tea I would use Orange Pekoe; and I would buy flea collars for the cats because I could never keep it straight in my mind whether pennyroyal was good for tea or bad for fleas.

I have mentioned the solitude of the first autumn on the farm yet I have not been entirely forthcoming. As I settled into the house, discovering with agreeable surprise that the plumbing plumbed, the well drew water, and that it was all beginning to feel remarkably like home, I noticed that certain other creatures were sharing it with me and that they were not all crickets. In point of fact, they had been in the vicinity the whole time—ever since August when my daughter had sold her college books, packed up her macramé and the *Diaries* of Anaïs Nin, and set off for Ibiza to join Itsy and the Rainbow Family. She had thoughtfully left behind, pinned to the wall above the kitchen sink, a note asking me to please water her alfalfa sprouts; and she had also entrusted me with a number of domestic friends, both four-footed and finned—not to mention the late Fester with his case of chronic root-bind.

Counting them over, I discovered that the total company in addition to myself included the dogs Booey and Alice, the Siamese cats Ferd and Fanny Farkle, and the goldfish Igor and Boris.

These of course were only the beginning, for a farm draws freeloaders the way an artistically gifted ant draws honey. Solitude, I thought, swiping at the back porch with a broom and watching them scatter. In a pig's eye. Except for there being no people, it was just poetic license.

# Chapter IX

# Farkling in the Night

It hardly seems worth mentioning that all I had really wanted for the farm was a trained Chinese goose.

This remarkable fowl, according to *Sunset* magazine, which panders shamelessly to Western suburbanite phantasies, would weed orchards, "or any place where plants are tall so the goose feet can't squash them." And they made excellent watchdogs. It was likely that a trained Chinese goose would be a "first" for the canyon, bringing untold status and honor to the owner. However, as I say, it seems hardly worth mentioning since *Sunset's* editors, obviously feeling that being first-with-the-news discharged their responsibility to all but the most unreasonable reader, failed to mention where one might get such a rare bird.

I put out feelers, only to have a curtain of chilling silence drop over the whole Chinese goose picture. Clarence, at the feedstore, hinted that he had troubles enough with his malady without making the CIA's Man of the Year list. My long distance plea to United Nations headquarters proved equally frustrating: the People's Republic delegation were coolly fielding all inquiries to a number identified as the Kowloon Konnection. When I dialed this, I got only a recording in Mandarin that was incomprehensible to me except for frequent references to something called Chuck's Duckburgers.

Well, I suppose I might have hung in there, learned pingpong, and in time have penetrated the Red Goose Curtain; but what with one thing and another, mainly Booey, the matter completely slipped my mind. Even I came close to slipping my mind. As for Booey, who had no great mind to worry about, he was simply concentrating with fearful intensity on slipping his chain.

He was called Booey in recognition of the natural bandit's mask surrounding his yellow eyes plus the whimsical pretense of passing as a watchdog. He not only looked like an outlaw but he was in fact a known lamb-killer from a ranch in Sonoma county. When he arrived at Meanwhile at the age of six months, he weighed in at one hundred and fifty pounds.

"What on earth is it?" I asked.

"St. Bernard, German Shepherd, and red wolf," said my daughter.

"I thought wolves were gray."

"It's a very special breed, Mom. The red wolves live in the Southwestern Desert, in Texas, I think. You might say we're helping to preserve an endangered species. And in this particular *unique* combination—well, he's probably one of a kind."

"Yeah. Well, I'll say one thing—he's no trained Chinese goose."

Booey, by way of saying Hi!, rose up and placed his paws confidingly on my shoulders, wagging his tail vigorously. A remarkably untrustworthy light flickered in his eyes. It was accentuated by the fangs in his huge head which, in my imagination, were flecked with gouts of fresh sheep's blood. I had no doubt that he could wipe out a flock of prize Merinos in a matter of minutes. And from that moment on I lived in terror of his reputation leaking out among our local ranchers, one of whom owned just such a flock.

He had extremely long legs, immense strength, and the fleetness of a springbok. Running in the hills he loved above all things in life. He always ran about twenty miles before breakfast and at least twice that afterward—*if* he had managed to either slip or break his chain, which he did repeatedly with the skill of a Houdini. Every link in turn became the weakest one. The strongest snaps were themselves snapped. Nothing held Booey.

As a farm dog he was slack and irresponsible; as a watchdog, seldom home except at mealtimes. The idea that I might ever need him was so remote as to border on the hysterical; yet, unbelievably, I found it hard to resist his guile and raffish charm.

The even younger Alice ran with him in the hills, keeping up as best she could. But whenever he slowed down and approached her at a slouch with a certain leer in his yellow orbs, she promptly sat down. She would remain glued to the ground until his attention wavered back to running, which normally happened in a matter of seconds. I attribute Alice's undoubtedly superior intelligence as an adult to the fact that she spent so much of her puppyhood sitting and mulling things over.

The bad 'un made it clear 1) that I was the supreme object of his worship and that he only awaited a worthy opportunity to lay down his humble/noble life for me and 2) that he hadn't the least intention of ever obeying a single order I gave him. For a time I engaged a dog trainer to give him obedience lessons, a conceit that in retrospect seems laughable. I suppose the trainer concluded that keeping his arms in their sockets was more important than teaching Booey to heel, a decision he reached in very short order. In fact, the poor man dropped out without notice, even failing to return for his final pay. When I tried to telephone him, I was told the phone had been disconnected. The only thing Booey had learned was to turn promptly, on hearing his name called, and vanish in the opposite direction.

I decided that Lorenz was probably right in concluding that male dogs of the wolf breed (or mix) are incapable of accepting orders from women, since to them any female is just one of the pack—although in Booey's case, the male dog trainer had seemed equally defeated.

On walks I could scarcely speak to Alice or motion her to stay at my side without Booey's huge bandit's head thrusting us apart. The only reason he ever walked at my heel was to keep her away from it.

When not engaged in chasing deer or in pup-molesting, he usually spent his time out in the field, bounding up and down on his staglike legs in unconscious mimicry of the tiny bouncing creatures he pursued, the grasshoppers, rabbits, and frogs. It was an absurd spectacle. Sometimes he tiptoed solemnly around after the little creatures, trying to take them by surprise—which was even more weird to watch. He would have looked more at home on a Harley-Davidson.

Alice often turned to me from her prudent posture and exchanged a glance. We enjoyed almost total communication.

In the early months at the swamp I learned a new (to me, at least) bit of canine lore. Dogs bark, as a rule, not because they are naturally aggressive and want to attack whatever might have excited them, nor because they want to protect their owners, but simply because they feel threatened. When the two dogs slept outdoors for lack of a kennel, they barked at every noise. They and I were exhausted from sleeplessness by the time the dog run was built. As if knowing it was for them, they entered it eagerly and with patent relief. From then on the nights were wonderfully quiet. They would have slept through anything.

Ferd and Fanny Farkle, the Siamese siblings, were patrician and beautiful. They were also possessive, jealous, vengeful, vain, hypochondriacal, calculating, and more than a trifle mad. As kittens, they had been pushed on us in a cardboard box by an inscrutable teenager lounging against the front of the Consumers' Cooperative in Berkeley. They were city cats to the core. Their Berkeley days thereafter were spent in pleasant degeneracy. Ferd hung around with his extended family at a commune down the street while Fanny, having ingratiated herself with an elderly nuclear couple next door, lolled on a claret-colored needlepoint love-seat through the long afternoons. But they were always at the front door to berate me with hoarse charges of neglect when I returned from work.

Nothing could have appalled them more than our move to the country. Everything about rural living was intensely repugnant to the Farkles: its rudeness, the lack of an intellectual hub, poor reception on the PBS channel, and the absence of any substitute for Ferd's unsavory group. Most Siamese cats are con artists at getting themselves adopted in their neighborhoods by pretending to be mistreated or misunderstood at home. One handicap of living in the country was that Ferd, who tended to imagine enemies lurking behind every blade of grass, was afraid to walk to the nearest farm and offer himself for adoption. He simply went into withdrawal and a terrible decline.

Never for a moment did the Farkles let me forget that I alone was responsible for their fate and that if I were not held to account for my sin in this world, some vengeful Siamese cat-god was sure to catch up with me in the next. There were rumblings that my lapse might even be (in Fanny's word) "actionable." I offered her a change of venue—to the pound—but she pretended not to hear.

The goldfish Igor and Boris had the grace to succumb to a case of bends the first time I changed their water. But it looked very much as if I was stuck with the feline and canine contingents.

The fates, however, soon intervened. On a dark evening in October a large truck hit Booey, killing him instantly. It could as easily have been the other way around. Only later did it occur to me, with horror, that I was uninsured against a dog demolishing a truck. In any event, his death, cutting short as I had no doubt an impressive criminal career, came as a terrible blow to us all.

Alice awoke at first light the following morning, discovered her aloneness, and uttered a chilling, wavering wail of bereavement, "AaaaoooOOOooo—oooOO?" It was a cry I have never heard her make since.

I got up and let her into the kitchen. She bounced twenty times in one second in tight circles and flopped down, not waiting to be told. She did everything quickly, like a trained sheepdog. The Farkles were glaring down, dark-faced and blue-eyed, one from the kitchen wall divider and one from the top of a dish cupboard. As far as they were concerned, letting a dog into the house was the beginning of the end, the downfall of empire.

I fed Alice. I boiled water and made instant coffee. The Farkles had been noisy in the night and had cleaned up their food. I put more into their plates, hoping to buy time, and carried my coffee back to bed.

The owl that had hooted in the night was still hooting, rather close to the house now. A solacing, warm, and lovely thing, the hooting of an owl.

I was balancing the hot coffee on my chest when Ferd crashed down from the open loft onto my bed. He glared at the puppy lying close by. Then, snuggling up, he curled a possessive paw around my neck and sighed. I could not recall ever having heard a cat sigh before but, like the hooting of the owl, the sound was comforting. Everything Ferd did was calculated to soften up the intended victim.

The moment Booey died, an abrupt dislocation occurred in the whole domestic power structure, with Ferd attempting (and pulling off) a takeover that left Alice permanently shaken and highstrung. When my friend Joe, a psychiatric social worker in New York, came to visit, he mistook her affliction for what he glibly labelled R.A.—or rape anxiety. That might have been true when Booey was alive, but after she fell under the Machiavellian spell of Ferd Farkle, her nerves began to shred very quickly. She got confused about things like horses and bulldozers. Sometimes when I opened the front door, she would rush out, bucking, with her hackles raised, crying, Wuf-ff! And there would be nothing there—just nothing.

Ferd's game plan, as the next few weeks disclosed, was to spend his days on a purple velour pillow on a little bed in the sleeping loft, from which he was able to pull the strings without fear of reprisal. When not sleeping, he laid schemes for psyching out the other members of our little group, employing an intricacy of plot that would have done credit to a Borgia. As the whim struck or as we below grew careless, he softened us up by ambushing us with sudden, leaden drops. It is a well-known fact that all domestic animals can make themselves light or heavy at will. Rufus, our old Labrador, who had weighed eighty pounds or so, could ease himself onto the bed at night as lightly as a leaf. So could Ferd, if he wished, but he chose to weigh in at around one hundred pounds when dropping straight down near Alice.

She, who now slept on a rug near my bed, was soon reduced to a quivering mass. Ferd would pounce, whirl to glare for a terrifying moment straight into her eyes, and then streak to safety.

For my part, I never dozed off at night without a precautionary look overhead. Usually I found Ferd's face peering thoughtfully down. On being observed he would glance quickly away. Not until I was sound asleep did the arch-fiend strike.

I suppose most people would have killed the Farkles for the tricks they pulled, but I felt rather sorry for them, knowing how they pined for the evils of the city. And they, of course banked heavily on my sense of guilt.

Ferd was notorious for what psychiatrists call "acting out." Usually one didn't even have to have a situation for him to start over-dramatizing it. He was a daydreamer, fantast, hypochondriac, and clinger—in other words, male all through. In his murky, private scenario, the house was always full of enemies. He was in constant rehearsal for the lead in "FBI." He would stalk around with his shoulders haunched and his tail bushed out, peering into every closet, growling, and scaring himself badly—when the only other cat within half a mile was his sister Fanny. Or he pretended that timid Alice was Organized Crime.

Just by looking at Ferd, you knew that eventually he was going to have trouble with his prostate gland. Actually, as it turned out, he developed cystitis on a weekend when I was in Southern California. I returned to find my bed pitifully flecked with blood. Ferd lost no time in letting me know that he was the emergency case and that we must hasten to the nearest veterinary, which of course we did. Always. Other times, feeling slighted, he would simply retire to the loft and vomit quite noisily. When I went to investigate, I would surprise him peering confidently over the ledge.

As for the fragile, crosseyed, little Fanny, she somehow managed to give the impression of going through life with her hands on her hips. At the farm, to everyone's astonishment, she developed into a vicious ratter. My neighbors, plagued with rodents and with only lumbering field cats on duty, envied me her hunting expertise and relentlessness. And because it was the first time that Fanny had ever excelled at anything, it went to her head a little.

One curious result of her new outdoor life was turning square corners. In the fields she felt safer following fence-lines, a precaution that she began employing when in the house. Walking across the living room, she would suddenly turn sharply at a forty-five-degree angle and head on a dotted line for the kitchen as if it were the only way to get from here to there.

Ferd soon had her whiteslaved into bringing him a moveable feast of mice, shrews, voles, moles, gophers, and kangaroo rats. When he heard the little creature staggering through the cat window with her night's catch, he would arise, stretch, bare his clean pink mouth in yawn, and toddle down for inspection. The offering, if it met with his standards, would be carried to the bathtub which he found a convenient abbatoir. It was part of the Farkle master plan that I should arise each morning to be confronted in my tub with a small bile sac or two and the grisly remnants of snouts, whiskers, and tails.

Occasionally Ferd (Fagan) Farkle ventured outdoors with criminal intent. But, as might be expected, he specialized, catching only child-mice up to about one inch in length, which were to be found with little difficulty around the woodpile near the house. After scaring them to death, he would place their tiny corpses behind a chair in my bedroom, and would then retire to his aerie to await reactions.

Of the two, Fanny was the more vindictive. One cold November morning at around 4 o'clock, she roused me ceremoniously by peeing on my head. Needless to say, this caused me to sit up quickly, wondering in a bewildered, sleepy way what I might have done to merit such exceptional recognition.

The only clear fact occurring to me in my daze was that it was the second time I had been singled out for this attention—both times by Siamese cats, although years apart. Why? Why me? Was it perhaps a case of other people concealing having had their heads peed on by Siamese cats? More disturbingly, was it this characteristic that subconsciously attracted me to them, a sublimated pee-wish, so to speak? In the blackness of my ruminations, anything seemed possible.

What had I done? How had I failed? Investigation disclosed that Fanny, with her habit of thinking jurisprudently, had three probable causes of action.

I had inadvertently let a door swing closed that excluded the cats from access to their little window and the outdoors. I had allowed their sandbox in the kitchen (Ferd, of course, requiring indoor amenities) to become somewhat less than dainty, a point on which the Farkles were inexorable. To cap it all, I had forgotten to put food in their dish although it was well known throughout the universe that all Farkles wanted a noisy snack at around 4:00 A.M. I wondered if it were too early in the morning to call the Public Defender.

After my passions had subsided under the effects of an icy shampoo in the usual ratty bathtub and I had crept shivering back to bed, Ferd dropped down at only about fifty lbs. to make amends for his sister's lapse. He purred and rubbed his whiskers against my cheek. He snuggled a velvety black paw around my neck. I soon began to feel an unpleasant irritation but put it down to the movement of blood back into my head. When I next awoke, I found not only that I had a head cold but that I had been given poison oak.

Yet another disgusting Farkle tactic, if I forgot to change their sandbox, would come to my attention as soon as I leaned over the basin to brush my teeth. The unmistakable odor was a message more eloquent than a thousand pictures: "Just a friendly reminder, F.F."

Well, I had never gotten off easily with cats. Take Alfie. The habits of the outrageous Farkle siblings often caused me to remember with affection this somewhat retarded lilac Siamese who had preceded them in our household at Berkeley and who had taught himself to use the flush toilet for micturation.

Because that had been the only triumph in Alfie's life, I had praised him highly the first time I noticed. From that day forward, he was precocious and punctual in his morning dash for the bathroom. It got to the point where he was consciously racing me for the privilege and winning. I retaliated by putting him out through the kitchen door first thing in the morning and then opening a sliding door on the sundeck, with the thought that Alfie, after patronizing the garden in a leisurely way like a normal cat, would be able to proceed around the house and reenter it at his own pace—which, in a sense, he did. He literally flew around the house, up the stairs to the deck, into the living room; and by the time I got to the bathroom, there he would be, perched triumphantly on the toilet seat, a quiet smile hovering just above his recessive chin. At that point I began to worry less about his mental retardation and more about my own. Nothing made a woman feel much sillier than waiting for a cat to finish in the bathroom. I never found a satisfactory way of explaining it to house guests. But at least with Alfie—who, to my sadness, passed on—I had never feared a malicious assault on my head.

In time two other cats joined the Meanwhile cast: Abigail, a huge gray field cat, and Edna, a young, stray, brown tabby whom we christened—after observing her habits, which included bringing home live rodents and keeping them in boots—the Dreaded Edna. Abigail was a ringer for the Cheshire cat in *Alice*, except that instead of grinning constantly she spent her days on top of an eight-foot wardrobe wearing the expression of a morose owl. She had good reason: every time she came down, the Farkles beat up on her. Finally we found a good home in town for her. Edna, although only an emaciated waif when she arrived at the back door, had the wit and bravado to put the Farkles immediately on the defensive. As a result the three of them became good friends.

The pup Alice, until she was about four months of age, had been an undifferentiated ball of brown wool exuding a sour, mashlike odor. (My daughter had said, "I used to think that puppies smelled like pet stores; but now I know it's the other way around.") None of us who knew Alice in infancy held out much hope for her personal beauty. But everyone agreed that she had lovely, bright, brown eyes which, like Grishkin's, were underlined for emphasis.

Suddenly, after Booey's demise and the new responsibilities thrust upon her, she blossomed out into a beautiful mutt, mainly Golden Retriever. She developed a fine, waving plume, a deep chest, and big feet with thick winter hair sprouting in tufts between the toes as befitted a No. 1 farm dog. She settled into her new job with such zest and enthusiasm that I guiltily realized how deprived she had been all along.

Bright she certainly was, and quick to sense moods—not only mine but those of visitors and of creatures she met in the fields. With dogs, cats, horses, cattle, chickens, ducks, and so on, she showed a rare appreciation of their "panic distances," dropping of her own volition if the animal or bird in question were ready to bolt, and would remain motionless until the situation sorted itself out.

Although the Farkles terrorized her indoors, she sometimes chased them in short bursts outside just on principle. Even so, she was careful to give Ferd enough leadtime so that his massive sense of self was not shattered—merely outraged. I learned that animals and birds are assertive on their own territory, but give ground on that of another creature, quite as a human being defers when out of her own depth. Alice, for example, would snap at our ducks when they hung around the yard, but when she stood belly-deep in their pond, intent on frogs, the ducks would swim up and peck at her.

She worked patiently to cultivate some very curious interspecies friendships. One was with a male duck that started coming up to the kitchen door with its mate and presumptuously challenging Alice on her own territory. When it attacked she would snap back at it, merely grazing its feathers. But one morning when the drake attacked, I saw her retreat as if in fear to the steps. When the drake reached her, however, he began preening her fur with the same courtship gestures he used toward the duck. Alice sat there quivering with pleasure, and I got the strong impression that she had planned this move. From then on, they spent quite a lot of time sitting side-by-side on the steps, staring straight ahead with sappy expressions.

Next she tried to strike up a relationship with two horses in a nearby pasture, where they stood most mornings in a proprietary way around their bathtub. Alice would go over and lie down quite close to them. Nothing much ever seemed to happen. Once I saw Alice make so bold as to sniff at a horse's hoof. The horse snorted in an arrogant, unfriendly way and walked off. But Alice kept going back every morning and sometimes she brought home a lump of horse manure for her collection. I felt she was hoping the horses would chase her; yet she knew that, emotionally, they were not ready for it. One day one of the horses urinated and Alice arose from her patient vigil and did likewise. It seemed to me that she did just about everything a dog could politely do to extend paws across the border, but the horses weren't having any of it. After about two months, she stopped going to their pasture.

Fanny Farkle became Alice's close friend. She enjoyed standing under the dog's stomach, complaining loudly while Alice pretended to swallow the cat's head and then make it reappear. Ferd Farkle avoided these disgusting performances.

Alice was actually a traitor to her own species. From her quickness to learn commands, I realized that Booey and the other male dogs we had owned in the past must also have understood a bit more than they cared to let on. Obviously, she had been absorbing the dog trainer's words while everyone's eyes were on Booey, because she picked up literally in a day or two at around the age of six months my commands to "stay" and "heel," that many dogs seemed to learn after a year or so of constant nagging. Obviously she had been starved for the opportunity to please a human being. In training this is half the battle. Since I dislike seeing animals made to perform like little Nazis to gratify their owners' power trips, I insisted only on enough obedience to prevent her from becoming a nuisance or getting killed on the road.

In short, I quickly spoiled her, with the result that she soon started second-guessing me on commands, and like a child testing whether I really meant what I said. If I gave her a careless order that made no sense in view of what she knew I was going to do next, she would quietly ignore me and do it her own way. If, for example, I left the barrow out in the field, and started back toward the house, absently calling to her, she would just sigh and flop down by the wheelbarrow.

She was the only creature of another species I had ever suspected of having extrasensory perception. Yet, when I thought about it, perhaps ESP was more natural to less evolved creatures than to human beings. If I thought about going for a walk, she would jump up and shake out her ears.

One morning I went to get grain for the ducks and chickens from a storage bin and found that the cut-off plastic bottle that I usually used for a scoop was missing. I started ladling the grain into a bucket with my hands, irritably wondering where I had left the scoop. Just then the farm dog came racing up with it in her mouth, having noticed that I had left it on the porch.

Sometimes when Alice's bright brown eyes were beamed my way, I would remember what Loren Eiseley wrote about coming across a frog's eye low in the water, warily ogling the shore, and how he paused to avoid frightening the creature. The frog's eye made him reflect on the inevitable surprise that one of our great observatory telescopes is bound to encounter one day while probing outer space. At such times, Eiseley says, he is reminded that "the most enormous extension of which life is capable" is the projection of itself into other lives. I would extend the "extension" to projections across the boundary lines of the species. For some human beings it is a short trip but for others frighteningly great, causing us to over-use such words as *anthropomorphize*.

If a bulldozer or a pickup truck tried to sneak into the yard, or if Morris, the neighbor's peacock, flew into a nearby tree and began sneering at her, fierce Alice charged them in her curious bucking way with her hackles raised and her beautiful tail wagging corkscrew style. Usually the invader, whether animal or machine, was so unnerved by the spectacle that it fled.

After Booey's death and her promotion to No. 1, she brought home a field worker's straw hat with chin straps and flung it ostentatiously onto her treasure cache near the small weeping willow tree. The cache was moved from time to time as prudence dictated. It had to be close enough to the house so that, when I started thinking about an unscheduled walk, she could dash out and grab her hat or some other special object for carrying and tossing, without the danger of being left behind.

Bones she quickly classified into three categories: cooked ones, for immediate eating, raw ones of no special aesthetic merit, for burying in the swamp with her nose and forgetting; and the special collector's items about which too much could not be said—her deer's shinbone with a dainty, polished hoof and a bit of velvet hide still attached, her acorn woodpecker, or her dessicated mole. Such prizes were taken on walks, tossed and caught, admired, and cherished. She might pretend to chew on them but it was only show, and in time they disintegrated from sheer over-concern.

The only times she failed to respond to my call were when she was absorbed in pond affairs, the frogs and carp being her special delight. After horses turned out to be such a disappointment to her, she spent hours lying quietly on the bank of the little island, or standing motionless in the water, admiring the light, the movement, and the water creatures. We christened the islet Alice Island.

Indoors, she demonstrated almost eerily noticing ways, such as never chewing on anything except during times of desperate need, when she would filch some cast-off out of a wastebasket.

When I worked in the garden, she always dropped everything and hurried over to lend a paw. If I were digging, she would conclude that I was trying to catch gophers, which led to a series of logical deductions—that I was a yard off the mark, that I lacked know-how, and that I seriously needed expert guidance. She would begin digging her demonstration tunnel next to my planting hole. Her dirt was soon flying into my dig, filling it up as fast as I could empty it, and into my shoes and socks. At intervals she would thrust her entire head into her excavation, snort around, pull it out covered with dirt, and look at me brightly. Obviously I ought to follow suit and improve my luck.

Or she might just pause in her project and watch mine for a while, giving a friendly nod of encouragement or a smile, but never criticizing or carping about my muddling, wrongheaded, countereffective approach. If the ground were especially hard, I could sometimes persuade her to dig a planting hole for me; but she always made it clear that I had chosen a most unlikely spot and that she dug there only to humor me.

When autumnal blasts whirled the leaves from the trees and swept them in golden drifts along the fencelines and the world thrummed with the excitement of migrations and last-minute food storage, I was inspired to action. I located a huge plastic sheet and began dragging leaves around to the front yard to mulch flower beds. Alice quickly grasped this maneuver. She picked up a rusty, leatherish sycamore leaf as large as a platter, carried it to the approximate center of the sheet and dropped it. Then she plonked her body down upon it and refused to budge until I dragged her off. When I began to pull the heavy mound around the house, she rushed to the other side of the plastic, seized a corner of it in her teeth and marched along—far enough to get me over the worst of it.

Here, there, to, fro, wherever I turned, she was beside me, or she was bucking along ahead of the wheelbarrow, pretending to be a horse. Thus by imperceptible degrees does a human being become enslaved to a domestic animal, until suddenly one discovers that one is hooked.

I found myself always in a hurry to rise on the farm—a compulsion shared by the dog. If the moon were brilliant over the meadow and the inside of the house washed in light, we sometimes went for walks at 4:30 A.M. Usually I let Alice out of her run at dawn. She would hurry off to the field to check several jobs at once, often arriving while late-running gophers were still flipping tiny showers of dirt into the air. Arrogance on the part of rodents always enraged her. Once she actually caught one and carried it home, fat and limp. After patent anxiety, she decided it belonged in Category 2 of Bones; but having buried it, she became very disturbed, returned to the field, exhumed it, and placed it on her special No. 1 treasure heap for permanent show. Next to her hat.

When I followed Alice into the meadow at dawn, I never knew what to expect. A leaf might fall or it might be in the very act of turning. A mother waterbug might be rowing her two oars across the pond to rescue an infant, both of them hysterical. We might see the indentation in mud of a small, sharply pointed hoof, the track of a creature poised in flight. There might be frost on the clay with, in half-light, Alice's coat a bronze glint; Alice's waving tail vanishing over the bank; and then, only a misty wreath of dog's breath hanging in the air, while underfoot the frozen weeds squeaked.

We might turn toward the hills where the sycamore trees had leaves like messy hair, like a high-water mark, the smallest always growing at the high tips of the branches and resembling, against a clear dawn sky, five-pointed stars sketched with a fine steel pen, but with the large leaves growing dense and messy in near the trunk. The heavier leaves in autumn formed heaps of rust-green-yellow tones that absorbed the outline of the dog. Alice, always searching, listening, her muzzle often bearded with dirt, her bright, myopic eyes alert to gopher trickery yet not unfriendly; ready to come to terms with man or beast; tail signalling good will and come-let-us-reason-together—even with horses. While I, still feeling a stranger in this land, tagged along.

# Chapter X

# On Rattling

Odd how one still rattled around on the landscape. The accustomed frame, both physical and psychological, had melted away, leaving only space as disturbingly unbounded as a canvas of Georgia O'Keeffe's.

Nor was it just the lack of job-identity, the professional anonymity, that might be expected to follow in some degree a drastic change, although this was rather more acute than I had expected. Such withdrawal pangs, I concluded, exist in direct proportion to a paucity of imagination and an excess of egocentricity. The sooner risen above the better.

The new expansion of my space-frame meant that self-scrutiny became painfully unavoidable. It was a time for flushing out the mind, clearing it of all that no longer fitted. Accretions of race wisdom oppressed me—tons of ill-digested, and to me, alien heritage. I rejected everything on principle—philosophy, prophecy, Holy Writ, institutions, In-Think, baseball standings, Jesus Christ, Edmund Wilson. Stand by for the big January Gray Sale. Everything Must Go—especially the Patriarchal Word. I should take my chances on risky, unfilled space and open borders.

That meant really being a loner, forever and henceforth a solitary person. To be a rejecter of doctrine and fad alike felt very solitary. The throwing out of excess baggage was the risky part. Only what I had learned by direct observation and direct intuition could stay. What if nothing remained?

I now understood all who raged calmly, having learned to live with their terminal disease. There was a time for everything: a time for the serene and incandescent state urged by Woolf for Shakespeare's sister's sisters; and a time for protest and whining and preachment. Then a third time remained—for calm, cold, terminal rage. One might hope for periods of remission, but who could say?

In our time we were lucky, for no moral leadership existed except within each individual. Thrust back on our own strength, we might learn what civilization could be. The rule of law could never amount to much without a spirit of anarchic self-leadership behind it. I used a made-up word, *autarchy*, to signify self-rule of the individual, because the others that might have done had been debased. An *autocrat*, for example, should be a self-ruler, but the dictionary defines him as a despot. How the debasement of ordinary language presages a deeper sickness.

After thrashing and rattling on my open landscape, more or less girded for a firm, moral stand against every sort of injustice, I found after a time, and to my surprise, that my passions began to suffer from a lack of focus. This was not simply because the war was drawing to an end and an age of dissent fading into history. It had to do with the country. A rural setting provides no sharp center, no bull's-eye.

Certainly, I thought, this calm will pass. This remission of anger is not right. What a fate—to have become a renegade only to lose my fire.

In any event, more pleasurably, I began to think about the kinds of people who were right for a farm and the kinds who might enjoy living on a farm but probably would not be right for it. The difference might be as simple but basic a matter as whether one were a morning person or a p.m. person. Animals awoke at dawn and needed attending to. At certain seasons one needed to find one's satisfaction—if any—in almost unremitting toil; or both combined. The days were never long enough in summer if a living had to be coaxed from the soil. Coaxed was a foolish word. One never cajoled nature. More like it were such words as scraped and prayed for and sweated. People should know.

I, with my thinking ways still town-rooted, found farm work greatly relaxing, varied, fulfilling; but I grew restless without changes of scenery. Even serenity could become oppressive.

It puzzled me that country people could regard this pleasant labor as work. Anything I did outdoors seemed almost play. I decided the rule must be that the moment one got paid for something or made a profit from it, it became work. If I ever managed to clear a profit on the farm, the blisters that I prized would no longer be the fruits of my vigorous play.

I noted detachedly that my increasing hybridization as a creature with ties to city and country had nevertheless produced no kinship with that bland invention, the suburb.

I took stock of my own physical changes in the course of a few months. The identifying marks of the countrywoman are striations along the cutting edge of the forefingers (weeding), a certain scruffiness at the ankles and heels, and other stigmata that would strike panic to the hearts of those whose mercantile fortunes or spousely security are sunk in perpetuating the myth of feminine frailty. Despite my difficulty in finding virtue in outdoor work as work, I nevertheless experienced keen moral superiority just from limping in, filthy and exhausted, at the end of a day.

Energy, I learned, could be rationed and spread out so that in a longer period than it might in youth have taken to do something or other, one could do most of what was needed. I learned that, when confronted with defeat, there was either another day or another way. Soon enough I concluded that one ought to keep trying to get organized on a farm but never, for the reasons mentioned earlier, hope to catch up. The great accomplishment was learning not to be oppressed by one's shortcomings.

I learned to tolerate but not relish the almost perpetual presence of small stones in shoes and the feel of socks forever welded together by the tiny iron hooks of burs. The development of callouses can be helpful, but they are seldom in the right place at the right time. Perhaps the nicest discovery was that flaccid muscles soon snapped back. My strength increased tremendously—from my having difficulty at first in lifting a twenty-five-pound sack of Eggena to being able to hoist fifty pounds of Startena; and then on to even more glorious heights—. In my desk-work days I had twice been hospitalized for a back ailment and told that I would probably always suffer from a "weak back." Quite a few farmers do, as I have discovered. One simply learns to balance properly.

Countrywomen seemed to enjoy an easier sexual equality than their city sisters. One reason for this is simply that the former are expected to use and enjoy their muscles, which are no cause for shame. The other reason is that the female roles on a farm or ranch are of equal importance to the man's and usually interchangeable—at least in one direction.

One of my two or three policy-decisions at the start was to accept no favors from neighbors that I could not return in approximate kind. The hardest thing I have had to do—and the most unsuccessful after completion—was to figure out how to raise a twenty-pound mallet well above my head, let it drop forcefully and raise it again, enough times to drive a whole procession of eight-foot metal stakes into the ground for the Banty chicken pen, or possibly 8,493 times. I finally carried out the project with brilliance and dash by dragging a picnic table around the field and standing on top of it. This enabled me to poise the mallet with both hands about two inches above the top of each eight-foot stake and then allow it to crash with sufficient driving force. Wisdom came with experience. I learned, for example, that driving metal fence posts should be done in wet weather when the earth is soft. The project was a failure simply because no fence exists that Banty chickens cannot get over or under, once they put their cunning minds to it.

I felt inordinately pleased, even so, by such admittedly simpleminded achievements. They helped enable me to bypass much of the old caravan route from freeway to supermarket. Even so, I still sneaked over to the market for pasteurized milk and a few other items that had best not see the light of a revolutionary dawn. Thus even in seeking simplicity, one compounds mental agony, creating a whole set of moral questions that never existed in one's conforming, decadent days.

Chapter XI

# Future-Glow

However humble her land or deficient in pH, a farmer tends to see it as a preview of a more perfect spread, every prospect being radiant with future-glow. This is especially true the first year or so when the general scruffiness, bareness, and barrenness demand a proprietary imagination to sustain hope. Often it is just what is missing from reality that keeps the farmer going, a fact that for some reason is not grasped by the nonfarmer.

My first city callers, as I noticed at once, pathetically hanging on their every word, looked around and saw things no better than they were: the naked-looking little house set among hostile growth in an unpromising swamp. I had traded *this* for the old marketplace of ideas at the crossroads of the explosions of new knowledge?

The fact that a farmer's dream stretched interminably and would never be fulfilled was a crucial part of the bargain.

In a nursery where I was buying small trees that would not bear fruit for perhaps five years, I overheard a woman deciding not to buy a particular kind of vine on learning that it would have no blossoms until the second or third year.

"I might not even be alive by then," she said.

She was a young woman. Had she been suffering from some terminal disease, with perhaps a year to live, I felt certain she would have bought the vine and planted it quickly. To make long plans, lives seem to need a sense of urgency. In any event the end product of a plant is only one phase of the pleasure it provides. Planting and growing it is another. A third part is whatever one can make of it in poetry and metaphysics. And a fourth part is eating the fruit, inhaling the perfume, or testing the new wine, as the case may be.

Before I planted anything and while it was still the first bleak winter, I saw, as all new farmers do, the entirety in vivid colors. Scarlet runner beans, daphne, tangles of sweet-scented vines, evergreen and shade trees, avenues of flowering shrubs, a ground cover of rosemary (tiny blue blossoms on leaves as gray as lichen and as green as the sea), and thyme; and viburnum, lilac, honeysuckle, climbing roses, passion flowers, bittersweet, strawberries, cherries, clematis, figs, plumbago. Long before a cutting was ever planted, I saw myself snipping a sprig of forsythia with buds and bringing it into the house so that I could watch the tiny yellow flowers, always the first of the season, opening in a vase, as once I had done in an English February. In the frame of the future were walnuts, almonds, a modest fruit orchard, a few raspberries, blueberries, grapes. Down at the western end of the marsh where willow scrub grew, I saw the emergence of tall cottonwoods, horse chestnut, eucalyptus, and of wild rose; plants and trees to fight the advance of freeway carbon dioxide with their natural oxygen factories. And in the thick growth, quail, marsh wrens, and other birds, both native and migratory, would find cover.

In the bog that drained the upper field, small creatures might continue to live, despite the inevitable arrival of more people in the canyon and on the surrounding hills. Skunk's Castle, as I called a great mound of briars perhaps fifty feet in diameter and half that in height, which covered an abandoned well, seemed safe for a good long while for the little procession—mother skunk and four babies—that emerged in a straight line each spring. The coyotes that I had heard on sparkling cold nights, the owls, raccoons, opossums, foxes, and cottontail rabbits would continue to hunt or browse there in the evening. Both mud turtles and turtle doves would find a meeting place as they had for centuries, but a jollier, safer spot; and deer might come to drink with less fear of the hunter.

The pond in spring was bordered—in my private preview—with wild iris and daffodils. Weeping willows shaded its bank. In time the twisted hands of old Arthur Rackham willow stumps would yearn toward the sun. All, in time. The vine maple, the two small Douglas fir trees, the salmonberry vines, the salal—brought on the airplane from another state, just as the seeds of morning glories would be brought from Ibiza, the peppers from Italy, the garlic from Greece—carried secretively in pockets and purses, all addicted gardeners being conscienceless.

And just here where Alice kept vigil over baby bullfrogs, here a flock of white ducks, two gray geese and three mallards would crash down each morning with a wild beating of wings. In October a wounded snow goose from Northern Canada might limp in for rest and recuperation. Later a pair of whistling swans, drooping after their flight from the Far North, across tundra and city—.

Then—a day in soft spring. Alice, growling in play, makes one of her little rushes at the swans and is viciously repelled by the cob. His mate has settled onto her nest on Alice Island.

A year or two more—the *Pinot Noir* vines are producing heavily. It is champagne for breakfast every day, because it would be a dumb *Cold Comfort* kind of farm without a dream like that. Alice, herding the pigs to root for truffles under the oak trees, passes me with a casual nod as I wander out to check the flush of mushrooms in the abandoned lime-stone quarry where the temperature remains constant year-round. It is nearly time to get the flail and beat the grain, bleed the pig, bruise the mint, trudge to worship—and all those other violently amiable pursuits that I saw, as clearly as anything, on the screen depicting the hours and seasons of Meanwhile.

I was insatiable. The *iffier* one's reality, the more gilded the dream. Bee-song, honey-drip, sweet-smelling meadow grass, ant-toil, apple blossom, perfume of green pippins winter-stored in a cold tin shed, pig-grunt, chicken-song, bird-cry, sigh of egg-fall. At night with the world a'glimmer, cicadas and frogs raise waves of counterpoint, the elf owl chuckles off, and barn owls hoot from the orchard or fly low through the trees, causing country people in their stocking caps to dream of levitation. The great willow looms, stipple-leafed and ancient, with gnarled trunk, silver by moonlight, muted by day, the benefactor of small creatures real and unreal. Beneath the waterfall to the trill of water music, newts dance solemnly in pairs, often curtsying to their queen. Salamanders, their red eyes still sealed, themselves almost un-born, tiny fingers extended, sprawl vulnerably in the mud. They cannot dance of course, all salamanders being wallflowers, yet if they do not get a move on a fox will snap them up.

The sights, sounds, smells, and doings inspired by future-glow so filled me that I was shocked when a guest from the city, pointing into the field to a bare place where scaly hardpan surfaced, asked "What are those shiny-looking patches where even the weeds don't grow?" and her husband answered with a brittle chuckle, "The heartbreak of psoriasis."

"Wrong," I said, unconvincingly lighthearted. "The real estate agent said that is a rich vein of Formica."

A cold wind blew suddenly across the canyon floor. We hurried in-doors. Why couldn't they see?

Here there would never be a day when the light struck a single tree in quite the same way as it had on any other day, or when any tree had the same number of leaves in the same stage of unfolding or coloration; or when a single blade of grass or wisp of cloud or brown reed at water's edge or ripple on the pond looked quite the same as they had on any other day or would again, for nature abhorred above all things monotony. Nature celebrated diversity. *Man* wore uniforms, felt safer in them, joined clubs, felt safer in them, formed nations, felt safer in them; and by erecting false barriers created fears which in turn created real dangers. Whatever one might say about nature as a wastrel, she was a great pluralist. She would never build tediousness into the days or drain the color from them.

# Chapter XII

# Hard-Time Apple Dilemma

The First Thanksgiving arrived in a year of no planting, and no harvest with the exception of bushels of bills for drainage, building, drilling, etc., yet bringing with it an indefinable excitement that seemed to presage a change in the weather. I felt a stirring in the marrow of newly sensitized bones.

On the haul road a single enormous sycamore leaf clung to the barbed wire. A few days earlier I had seen the body of a sparrow that had been killed by a shrike or butcher bird impaled in just such a way. Now the butcher bird of seasons had speared this too-perfect leaf. The air was motionless, the sky gray and cold. But winter, that old shrike's assistant, held back.

Despite the lack of a harvest, my cofarmer Barbara arrived to work on our plans for Meanwhile and we made it an occasion for various kinds of culinary license.

Although short on cash, we happened to be long on locally grown golden delicious apples, a fact that resulted in the creation of a reversible dessert. That is, the cobbler that I christened Apple Dilemma (by careful flipping and another touch or two) became an upside-down cake which henceforth was known as the Meanwhile Crunch. The guest responded with her own originals centered on the excellent local produce. When the holiday had ended the oven of the new kitchen range that I had scarcely used was splattered, blackened, and for all time initiated. Also my Indian Summer lean look had passed. So all holidays should end.

We talked about the minor mistakes thus far made in structure and concept, largely because of inexperience, too much haste, or by cutting the wrong corners. We resolved that once she came to live at the farm, we would make our corrections and undertake all future projects slowly, carefully, and with the love they deserved. And this gave rise to another matter—of attitude. Meanwhile demanded positive new attitudes of us. We wanted to keep it as a place where friends could count on finding affirmation. And oddly enough, although it was a hunble place and we never spoke of these matters to guests, sensitive visitors had a habit of suddenly bursting out with exorbitant testimonials to the warmth, beauty, and love felt in our living room. It seemed to me that all vibrations were heightened or intensified in the country.

tion effected by Sylvia Porter's insulation. Alice on her rug looked smug.

On the night following Thanksgiving I awoke with the feeling, still, of a significant barometric event in the offing. The roof of the farmhouse had as yet to be tested by moisture. I was both curious and eager to hear this new-roof reaction to the season's first rain, and also a little anxious. But what awoke me was less a sound of raindrops on roof than a feeling of softness through the open window; less the patter of drops than the movement of earth granules parting to receive them.

It was also the sound of a swampful of earthworms rising to excrete their casts, those little dark mounds appearing after a rain that were said to be prized by organic gardeners for their nutrient content. Now, in the dark, I fretted about what earthworms did with their sewage problems between rainstorms. Some gardeners, I had read, zealously collected these rich, dark heaps and sold them to nurseries. Once I began thinking about earthworm casts, as I did most worrisomely that night, I began to suspect that they were not excreta at all but merely the little hills made by the emergence of their bodies—in which event the organics were probably full of beans. Yet, as to the hardy souls who collected casts and sold them, I could not help admiring them. I tried to imagine myself approaching a nursery owner with my small basket.

"Pardon me—!" No. Forget it. I could never pull it off.

Still, I pictured his searching glance.

"What have you got there, lady?"

"A fine rich lot of wormshit, sir."

"You've got to be kidding."

"Worth its weight in gold on today's market."

In the expectancy of the night I considered the unreliability of a cash crop that emerged only after rain, in a dry state. We were risk-takers, we land people.

Through the open window I sensed the movement of pocket gophers waking from curled sleep in the antechambers off main runways, where they had chewed and stored up huge balls of dried grass against emergencies; heard them rushing to their entrances to block them with earth. "Batten the hatches! The End of the Wo-o-orrld is Coming!" I imagined them stumbling over a welter of uninvited fieldmice, dormant gopher snakes, king snakes, rattlers, tarantulas, and God-knew what else that was snoring away down in those tunnels.

The sounds I heard were the collective sound of the swamp preparing for rain.

And then the droplets began to spatter upon the new roof. Spatter-spatter. Faster, and faster. Heavier. Patter-patter-patterpatterpatter. Terpatterpatterpat-terpat. And then from the living room, from one of the adobe tiles near the fireplace, *Splat!* And a repeated, dull *Splat-Splat!* I tuned my ears for leaks elsewhere in the house but heard nothing. Well, if that was the worst, I would have it fixed in a jiffy next morning by climbing onto the ridgepole with the left-over squeezings of Dabble, Dribble, and Squish. From the loft I heard a Ferd Farklish sigh of deep contentment.

The calming rain continued and I slept.

By morning the small storm had ended but mist was edging in through the canyon. Warm air and clouds of fleecy full gray. The earth a sponge. Although gardening books warned one not to dig in such weather because of the danger of compacting the soil, I could not resist this first chance to hack into ancient, unyielding earth.

And once begun outdoors I found it almost impossible to stop, especially since I could now feel the new blisters of moral supremacy blossoming on the palms of my hands. I doubt that any farmer ever loses completely that initial godlike thrill at the prospect of making things grow, not with spring coming around as regularly as it does.

The co-farmer volunteered to trespass upon neighboring grazing land to steal manure from beneath a massive oak tree where cattle stood for summer shade and for shelter during winter storms.

"Well-aged manure," she boasted. "Just what all the books insist on."

She wheeled her wheelbarrow of cow chips around to the front of the house. We both felt gladdened by this spot of larceny so desperately needed by our soil. And few things in life, I reflected, were as invigorating as breaking the law.

While I dug at the earth spread by Entwhistle in the circle in front of the house, readying it for three white alders, and employing a pickaxe even after the rain, Barbara continued to haul manure and heaps of soft, wet leaves for mulch.

"Here's for the Mr. and Mrs. Nieman Marcus rose," she cried, throwing a spadeful of cow chips.

"Here's for the Mr. and Mrs. Ronald Reagan yucca," I echoed, adding a spadeful. "And my Flora Norton harebell-blue sweetpeas."

"And my Dorothy Eckford white-toned variety sweetpeas. And my Mrs. Collier primrose-colored ones and my extremely delicate-scented Lord Nelson and—."

"This circle," I said, "may have to take care of the vegetable garden too. Because thus far no one around here has a tractor and disk. We'll have to wait and see."

"It's only November. Maybe the rain will soften up the earth by spring."

"Ha!"

One of the longtime successful farmers had brought us a bag of fat, juicy garlic bulbs. "Plant them," she had advised, "in the dark of the moon during December, with a screwdriver." Although I was jumping the season by a month, I tried in other respects to follow instructions.

"They don't seem to be screwing in very well," I complained after a while.

"Try this," said Barbara, pointing to the pickaxe.

I finally substituted a crowbar for the screwdriver and thus managed to inter our first crop.

"What you must do to repel the gophers," volunteered Barbara who, like the Countess, had read a book, "is to plant garlic and daffodil bulbs around among the trees and vegetables, or vice versa, instead of confining them to beds. Rodents won't bother around foul-smelling bulbs."

"I hope they have heard of it," I panted. "The way I hope the earthquakes have heard of the wholesomeness of these more or less continuous shakings-up."

Just that week several light quakes had occurred.

"I've learned something too on repelling pests," I added, leaning on my pickaxe. "The deer hate sweetcorn and won't touch anything with a stalk of corn growing in front of it. Another woman farmer disclosed this secret. It is surprising how many independent women farmers there are in this area—if the expression is not redundant. Anyhow, she had planted corn around her tomato vines and she showed me the hoof marks where a deer had gone right up to them and had then turned away. Only thing—I forgot to ask her how to keep the raccoons out of the sweetcorn so there would be enough of *it* to keep the deer out of the tomatoes. Why did I ever think this food-chain trip would be simple?"

Alice, unnoticed by us, had been staked out on a nearby gopher hole. Now she came to report that its occupant had been brazenly watching our gardening activities. (Probably sketching a small map. "X. Dig here.")

"Okay, Alice."

I dashed into the house and grabbed something I had just bought, an anti-gopher sulphur bomb that was supposed to asphyxiate rodents without causing them undue agony.

I lighted the wick and pushed the bomb into the tunnel. Then, in a panic, I pulled it out. It was shooting forth a long flame before the poison gas was released. A steel trap would have killed more mercifully.

For more than a year thereafter I let the gophers work their will, which seemed to be gobbling up everything in sight and the procreation of more gophers. I weakened and bought steel traps but they remained on a shelf unused. I bought some poisoned grain, and it stood beside the traps. Young trees I wrapped in shields of chicken wire before planting. Little Fanny Farkle patrolled every inch of the property in the relentless pursuit of rodents. I planted marigolds among the vegetables to discourage them. I rolled mothballs down their tunnels. I mixed oil into gasoline and poured this into their tunnels because someone said they hated the smell. Sometimes I punched holes in coffee cans, filled them with dirt, and planted new shrubs and bulbs in them. I suppose it may be said that the gophers and I reached an accommodation—for a time. But when large and healthy apple trees were chewed off at the base, I turned as nasty as a bombardier and took the poison grain off the shelf and stuffed it well down into their tunnels and I almost relished the thought of gophers with terminal bellyache.

But I don't know—we tried détente again. Which is to say that *I* turned for sustenance to meditation. *They* are still on young trees.

Late that after-Thanksgiving day the rain poured. Underfoot the soil had a sinking, bottomless feel to it that was quite disturbing when I tried it. Now I should find out whether the morass on which the house stood had been made safe by Entwhistle; whether the pond dam, made of an old redwood door I had salvaged from the county dump, was too high and might have to be pulled out in the midst of a flash flood— which I considered a distinct possibility on the little *ciénega* after prolonged and heavy rainfall.

From indoors I could see the deluge beginning to stack up, forming sheets of surface water. What would happen once every rodent hole in that earthy Swiss cheese was brimming?

But the storm soon ended. At early evening a fresh and beautiful mist came down over the hills, filtered into the canyon, and caused the contours to move forward as in an Oriental painting. Mist blurred all but the foreground of ancestral willow, the red haze of the creek dogwood branches, the green, thick bulk of the oaks.

Then before the sky had even begun to get dark, in the wintry soft dampness of these colors and the leaf-sweetness of the air, I heard the owl hooting his deep and reassuring voice. I felt glad to be alone again, glad to have the old owl friend out so soon. Now it was a comfort to be with just Alice and the cats in the wild country.

By dark the fog was touching the far end of the meadow and reaching along toward the house. The great Auden had said it simply, "Thank You, Fog."

Laura Raboff's box-painting stood at the western window, catching and impaling on plastic prisms the last rays of daylight. She had named it Star Ice Song No. 2, but I for good reason called it the Earthquake Machine. When tremors occurred, I could always tell by whether the little translucent slivers were dancing at the ends of their invisible strings beneath the silver-blue clouds of the box's interior if what I felt was actual—or whether just the hammertripping of my heart.

# Chapter XIII

# The Black Phoebe

In December I grew so attuned to the season that on gray, cold days my metabolism slowed like that of a raccoon in minihibernation or a yogi in a trance. I dreamt a curious dream of rootballs, lamb birth, and human birth growing in sequence from a single image. The rootball of the orchard seedling (my big planting job for January) spewed forth a bubble of the sort that lambs are born from. But when this bubble burst, a human infant stood there. It spoke to me.

I said rather sharply, "Babies don't talk." And I knew it ought not to be standing up.

This had to be a dream so I awoke. The curious thing the infant had said, enunciating clearly, was, "Thank you."

At the realization that I had begun to identify subconsciously and in such a grandiose way with the Earth Mother, I could not help giggling.

It was not unusual, as I knew, for women who had made drastic changes in their lives or who contemplated doing so to dream of having a baby. The desire for rebirth. Violet Staub de Laszlo, in her introduction to the writings of C. G. Jung, notes that "the dream figure of a child who is personally known to the dreamer, while drawing his attention to his relationship with this particular child, may and does nevertheless carry a symbolical message referring to the significance of the child image as the dreamer's potential for the inner growth of his own personality." More traditionally, the male psychiatrist interprets the dream as meaning that a woman wants to have a baby as a device to recapture a straying male. The infant in question was recognized by me. In strict translation, nothing could have pleased me more than to be assured of her gratitude for the gift of life. A mathematical genius, I reflected, once worked it out that the statistical improbability of any of us being born is so great—that particular sperm cell meeting just that egg, etc.—as to make it virtually impossible for anyone to exist.

It is time to mention the sad affair of the black phoebe.

This small bird of great beauty adopted our household during my first autumn at the swamp. Flying in the face of tradition, heedless of the rules, listening only to the vain and rebellious counsel of the heart, it failed to go South before cold weather struck. I became aware of it as a distinct and unusual bird in late October when the winter movements of most of the others had ended. Those who were going had gone. Those who planned to winter at the swamp had arrived, looked over the accommodations, deposited their key money, and settled in.

But between the last departure and the first arrival, on a desolate evening, I became aware of a low, flute-like *"Twee-eee?"* (The experts report that a phoebe is so named because its cry is *phoe-bee*.) Suddenly I realized that I had been hearing, without *hearing*, the mournful and imploring cry ringing out repeatedly across the fields and swamp. Now at last I took note of the fact that this tiny, wild creature had chosen to remain with us—and probably without knowing any more than I did of what the winter at its worst might bring. On cold mornings, noticing it fluttering pathetically against the eaves and windows, I mistakenly assumed that it wanted to come inside and get warm.

The reason a normal phoebe flies South before cold weather is that it cannot eat seeds but must survive on insects, preferably the flying kind that abound in warm climates. And the little black phoebe that stayed was fluttering against the eaves and windows of the farmhouse where, even on frosty mornings, a few insects could be found clinging to the warmish surfaces. This, alas, I did not learn in time.

As the months passed I became very attached to the resident bird with the black, rumpled brow—more than I realized. Its cry ringing huskily across the fields expressed something of my own sense of aloneness and the sadness of winter. Like other baroque instruments, it was part of the mood of the place. I even presumed that the phoebe might share my own perverseness, that like me it might be happiest in this cold, Slavic time.

Somehow it survived the cold months and remained on the farm through the following year. When the first spring came it did not, like more level-headed birds, take a mate, build a nest, and raise a family. Instead it just hung around the house, now and then saying *"Twee-eee?"* in a polite, inquiring way. If I sat in the backyard on a late, bright afternoon, surrounded by Alice, the cats, the Banties, and such, the phoebe would alight on an electric wire close to our group. Although its attention span was short and it often swooped off after a passing insect, the bird minded its manners and was civil.

But the second winter, during two or three weeks of December, the mercury sank to historic lows. Pipes, pond, and earth froze solid. Everything was ice-sculpted. Not a bug or a fly could be seen. I put out wild birdseed for the canyon seed-eaters and syrup for the humming-birds, which newspapers reported to be starving all over California. Each morning the sparrows, hermit thrushes, titmice, towhees, and jays would swarm to greet me and my largesse. Even the shy little lad-derbacked Nuttall's woodpeckers would come close to the house in search of comatose insects.

The black phoebe began swooping distractedly around the house, fluttering near the ground or against the panes, its mournful cries sounding increasingly desperate. It also began haunting the home of my nearest birdwatching neighbor who, undone by the sounds of hunger, told me that he had finally salvaged a few flies from an old flypaper strip and put them out, thinking that, although they contained nerve poison, the hungry bird might possibly not be susceptible to it. He reasoned that it was in any even doomed.

The following morning as I was on my way to the chicken house with a kettle of hot water to thaw the drinking bowl, it occurred to me, stupidly late, that a frozen heap of sawed willow chunks near the fence would have insects under them.

I kicked the chunks apart. A salamander moved slowly and con-fusedly toward an escape hole, feeling a terrible chill on its shoulders. An earthworm wriggled. Two immense black beetles bestirred them-selves. A number of fat pillbugs curled into perfect little round balls so as not to be mistaken for pillbugs.

The Nuttall's woodpecker, hungry for insects, was fluttering in my wake. I looked for the phoebe but it was nowhere to be seen. I listened for the familiar cry in vain. Each morning from then on during the freeze, I turned over old chunks of wood and sticks for the birds, berat-ing myself for not having thought of it sooner. It was not often in a life-time that a black phoebe adopted a human being. I had been honored. I had let the team down. And now I felt only an acute sense of loss. In subsequent autumns, I was surprised to find myself always listening for the phoebe's cry at evening, and always feeling a twinge of sadness when a bird call turned out to be from another bird.

During that winter of deep cold I learned something about the dur-ability of insects. Either they were capable of freezing and thawing out, or they found many hiding places in the bark of trees that protected them. For when the cold weather ended in rains, many flying insects were back in action immediately. The indoor crickets, a host of them that I had hoped were dead, shook off their torpor and began staggering up the walls again, looking curiously like babies just learning to creep. What I learned convinced me that the phoebe might well have survived had it been spared too much help.

During the big freeze I carried indoors for firewood some of the partly rotted chunks of willow wood that contained cells filled with frozen pillbugs, and thoughtlessly stored them in an old copper washboiler that served as a woodbox. The next morning thousands of thawed bugs were swarming across the floors and up the walls. As I vacuumed them up and swept the remains out the door, I thought again of the black phoebe for whom this might have made Christmas and many other dinners.

Country winters, even in California, were merciless to most creatures—the cattle, horses, and wild animals that failed to hibernate. Sometimes, like people, they got arthritis from exposure, but had only body language with which to convey word of their pain.

Chickens too, even if fully feathered, could freeze on the roost. A comb, if frostbitten, turned brown and should be trimmed off with scissors, according to my USDA pamphlet. I would go to the chickenhouse on cold, clear nights when I knew a freeze was coming, and push the young Banties closer together on the roost and scatter straw around them. Then I would hang old pieces of rug on the wall behind them. Even so in the mornings I would find them sitting on the roost with hunched shoulders, looking at the steam rising from their thawed drinking water, trying to decide whether it was worth the trouble to get out of bed.

For myself, I preferred putting on layers of clothing in the winter to keeping the temperature high in the house. Most canyon winters turned out to be three-sweater affairs. Sometimes they were also three-sock affairs and thermal-longdrawer affairs, and I would feel like the Abominable Snowman stumbling through drifts in the Himalayas. In summer, rather than using electrical fans, I preferred to reverse the clothing process.

In January Alice began her collection of ice art at the pond. She found that, by plonking heavily onto the ice with her large front paws, she could break out plates which she then carried up the bank. As always, her treasures were visually inspiring. One piece preserved in its fresh crystal a sprig of the hollow, segmented marsh grass with some green branches of watercress. Another held in its glaze two or three brown-rotted, finely etched maple leaves. These plates, ephemeral yet exquisite, were left on the frozen side of the field where they would last until we had spring rains.

I, in the winter stillness, had been liberated from hibernation. Now my brain churned and I read everything at hand.

Awoonor the Ghanian poet was despairful for his country in its early trials. But would his people be happier when they had more of the Western things that enslave as they liberate? Why could they not learn? A poet had to make trades that were not always good trades. An American poet might conceivably buy time, space, privacy, surrounding beauty, and enough but not a burden of material things. Yet I suspected that Awoonor had a sense of belonging to his people in their struggles, while the American poet would feel alien. American poets jumped from bridges, literally or metaphorically. We did not give them—any of our creative people this side of celebrity—a sense of being essential to the national pride as, for example, the French did. Even when things changed for Ghana, they might remain the same for the poet. Maybe a poet, being a poet, had the best of life already. And maybe that was not true at all . . . How strange it was, this winter aloneness, with no human sounding board.

As the early January cold felt about to break, I noticed that—as with human beings—no season was ever simply one season. Through being aware of subtle changes, anyone could fine-tune the seasons, creating more of them than the calendar provided. Long ago on a 12th day of August I had stood on the sunny shore at St. Ives, Cornwall, surrounded by pale Englishmen all holding newspapers over their heads, and at a certain moment I had smelled autumn arriving from across the Atlantic Ocean, autumn apparently delivered to me alone on the wind. Then again, on the Monterey Peninsula, a false spring came in the fall after the first rain and the paper narcissi opened. Now the trees in the canyon were so swollen with buds that they must ache, especially the red-branched creek dogwood and the willow; yet the big trees, the sycamores, that were late to leaf, had never been so revealing of their cold skeletons and for some of them, spring would not arrive until May. All growing things made their own calendars.

All of one day the clouds passed, a few flocks at a time, over the canyon. And at evening great formations, wonderfully graceful, of what were called mares' tails, the swirling precursors of storm, leaped over it. Then they became white egrets soaring.

Crossing the clerestory window was a many-legged bug that moved with the lethargy of winter, all its legs and feelers lacking coordination, a sad, comic bug beset by cold, probing here and there while making its way across the great Arctic tundra of my window. A crack between warm boards was its goal.

In this dead of winter I came again across the work of "Anon.," the poet whose lines reminded me of John Donne's and again I found it hard to believe that nothing survived except this poignant quatrain:

> "Western wind, when wilt thou blow,
> That the small rain down can rain?
> Christ, that my love were in my arms
> And I in my bed again!"

Old slogging, footsore varlet, what happened? Did you die on a battle-field, beneath your horse? Did your body of work fall into a moat? Or a prudish relative burn it? Perhaps the poet was a woman and her famous spouse plagiarized the rest of it. Because there had to have been more. Possibly the lines were secretly written by Queen Elizabeth at Tilburn in black August of 1588 when the Spanish Armada threatened. She swallowed the rest of it in her tent so that the ubiquitous enemies in court could not ferret out her weakness, and bethink she had not "the heart and stomach of a king, and of a king of England too, and think foul scorn that Parma or Spain, or any prince of Europe should dare to invade the borders of my realm . . ."

And only later I came on this when rereading Virginia Woolf: "Indeed, I would venture to guess that Anon, who wrote so many poems without signing them, was often a woman. . . . This may be true or it may be false—who can say?—but what is true in it, so it seemed to me, reviewing the story of Shakespeare's sister as I had made it, is that any woman born with a great gift in the sixteenth century would certainly have gone crazed, shot herself, or ended her days in some lonely cottage outside the village, half witch, half wizard, feared and mocked at."

On a morning in mid-January Alice and a neighbor's dog began barking at a raccoon weaving on its haunches on the frozen swamp. I chased the dogs home. Wanting to take food to the raccoon, I rationalized instead that by encouraging a hungry animal to hang around the farm in winter, I would quickly make it dependent and sooner or later there would be a bloody showdown with the dogs. I watched it from a little distance. It sat in a tilted, triangular shape, its head shaking.

In the house, my work went badly. Ferd Farkle began racing around, ricocheting off walls and furniture. In one pass, he careened off the side of my desk and in the next, bounced off the end of the In Basket, causing papers to spill gratifyingly. Next time through the room, he figured out a way to ski across the spilled papers in a long glissade. He then jumped up beside my electric typewriter and began rubbing his whiskers against my hands and wrists, causing the space bar to jerk. Tiring of this, he tried to catch the flying keys. Tiring of that, he overturned a jarful of pencils. I remembered some errands I had to do in town.

A few minutes later, driving out the gate, I noticed that the gray raccoon with the frosttipped fur was humped over, its head hanging down, a triangle from which the life was ebbing.

When I returned, it lay stiffening on the earth. Then, irrationally, I felt remorseful, wishing that instead of thinking about it I had simply brought it into the house, warming it, giving it food. Winters were cruel. Hard for the black phoebe, hard for the poet, hard for the range cattle, the raccoon, and for the old English Queen who, standing erect for hours as her death came, had said, "A weight of iron is around my neck."

Assuming that she *had* written the quatrain, I reflected, and had *not* chewed it up along with the rest of the body of her work, perhaps even so they would not have sent her to the Tower. But surely she could not have continued as Queen. It wanted a leveller head for that. The free spirit that elects to fly in the face of the system, or which—as in the case of the black phoebe—elects not to fly at all, must pay some toll. That was probably when they gave you the flies that made you laugh yourself to death.

# Chapter XIV

# A Special Breeze

Few Western towns near great population centers have been as blessed by oversight as San Solace, tucked away in its valley, unpretentious in its ties to the soil, and laying on no airs for tourists. The flat, rich fields with monster harvesting equipment and cubes of workers' camps stretch right up to the edges of town. Business streets bask in the dappled sun patterns of broad-leafed trees. Little mystery exists as to how folk earn their living (agriculture, the canneries, local business) or even of how they vote (two-thirds registered Democrats of whom many cross over to vote Republican if it looks as if a wild-hair Liberal or a Chicano might win a seat in Congress). A small version of what Californians mean in a collective, cultural way by the expression "valley town." No hurrying except when the crops are ripe—and then, man, watch out! Refrigerator trucks and trailers racing night and day, tomatoes all over the freeway ramps, sugar beets and kale flying through the air.

The town cherishes its flavor of the Wild West. Mayor Sam (Cool) Johnson, ace six-shooter, when interviewed on national television as to whether he had any advice for local citizens about a predicted earthquake, drawled, "Wal—only thing I'd suggest is—if they've got any good bourbon whiskey—that they put it in a safe place."

You could tell the racial mix from the kinds of bread in the excellent small bakery—Italian, Portuguese, Chicano, Anglo. There was also a sprinkling of Japanese farmers who seemed to have no traditional bread at the bakery, but you could see their little offerings of cookies if you looked on the beautiful tombstones in the Buddhist section of the cemetery.

Most are people with strong Old World ties to agriculture. The ethnic customs and formalities still survive right alongside the time-honored Yankee drive for the things that money can buy in Detroit and Grand Rapids. Longevity is common among those who survive their teens.

But the local fruit crops no longer compete very successfully with the big, conglomerate-owned orchards of the Central Valley, where harvests occur earlier in the season. More and more orchard land is being zoned for commercial uses. I feel glad to have known the town while it still has its country face.

A man called Will runs an unusual small shop which he crams with everything, so long as the quality is good, but mainly with the kind of clothing that country people wear. Dangles of this and that, boxes, bundles, all shoved in tight and to the casual eye, just anyhow—but this is deceptive. The winter when woolen mittens insanely were decreed unstylish and you couldn't find them in department stores, I went to Will's shop, with its shoe polish and Indian moccasins in the window and the nice 1957 calendar on the wall, and told him the kind and size of mittens I wanted. He thought a moment, without moving. Then he reached behind his back, fumbled for a second, and came out with a big box of them. What was more, the price was right.

People tend to develop a fierce loyalty to his shop, as if he carried rare and exotic brands. Actually it is just good practical clothing. If there are as many as two people crowded into the shop when you arrive, you can look for a seat on a liars' bench out front and maybe take care of some town politics. The big treat is to ask for jeans because then he leads you back through the family dining room, with its good smells and the samplers and religious pictures, and out another door into a large room stacked perilously with denims, woolens, thermal underwear, jackets, and caps. You have to keep your elbows against your sides as you move along the aisles. When I felt especially daring I would ask to use the fitting room, which meant negotiating corridors of leaning merchandise leading to a tiny cubicle in a corner. Will would switch on the bare light bulb and depart discreetly, leaving the customer to work out the logistics of a fitting room considerably smaller than a phone booth and furnished with a bit of mirror that gave you a grand view of your stomach.

At times he does a little leather work. Once he repaired some boots for me but wouldn't charge because the heels were hollow and could not be soundly fixed. I said, "What if we take off the caps and fill up the holes?" but he closed the matter with, "Wouldn't do any good." Another boot customer, who drives a chartered tour bus from New Jersey to California makes a point of bringing his passengers to San Solace each year for the simple, excellent reason that he wants to buy a pair of new boots from Will. "You can't *buy* leather boots," this partisan says, "on the East Coast."

The thing I most enjoy in San Solace, though, is just walking through the streets in the morning when garlic is being harvested in the nearby fields. Nothing is quite so restorative and invigorating as the clean, sharp whiff of fresh garlic on a morning breeze. Like bagpipe music, it is at its finest outdoors. But unlike bagpipe music, garlic fumes do not have to be receding into the distance for maximum enjoyment.

For anyone caught up in murky urban afflictions I heartily commend this pungent experience.

Smelling the wind permeated with fresh garlic has changed my entire attitude toward this vegetable in cooking. Where I once was fearful and sparing, I now find myself using it as if it were one of the aromatic seasoners as harmless as carrots, celery, or onions. It hangs on my kitchen wall in braids beside strings of red chili peppers that scorch the paint. ("Taste this soup and tell me what it needs." "More garlic!")

In the fields, workers and machines are engaged in an almost continuous tableau of harvesting. Crops appear to shoot up overnight, literally, and are dispatched in a frenzy. One morning you notice that the highway near the lettuce fields is lined with the sedans of Chicano families and see ancient, gaudily painted buses of the labor contractors parked in the fields. Mature heads of lettuce are automatically wrapped in slivered plastic and hurried into refrigerated trucks that speed them to grocery stores three thousand miles away in New York or Atlanta. Days later the same fields are being plowed again.

Crisp, flawless bell peppers are carefully picked by hand and sent off in tractor-drawn carriers. But tomatoes are harvested by machine, the workers standing near conveyor belts that waste portions of the crop and leaf the tangled vines splattered with crimson. The machines were developed at the University of California, Davis campus where plant geneticists later had to design a special, tough-skinned tomato to withstand the assault. They would go into cans for sauce, puree, and juice. Celery stalks are slashed by workers standing in rows, wielding their blades in the formalized rhythms of a dance. Huge flat-bed trucks laden with sugar beets rumble toward the railroad station. Other trucks race out of the vineyards, dribbling a thick purple stain on the dusty roadbed. A stranger might think that an army of wounded had passed that way.

Aerial vacuum tubes suck nuts from the trees, to be hustled off to a growers' cooperative for drying, grading, and shelling. Fruits such as pears, apricots, apples, and prunes are handpicked in hampers. Flower and vegetable seeds, grown with infinite care for new hybrid characteristics, are harvested, weighed into tiny packets, and passed on into the consumer market for next year's planting frenzy.

And then, as quickly as it all began, it ends. The ravaged fields, except for dying vines and rotting culls, lie bare while the earth, for a day or two, seems to pant.

But almost at once tractors and disks are racing across them again, turning the old vegetation back into the soil, preparing for another crop. On these commercial fields, not a weed, not a rodent, or an insect dare appear. The earth is subdued utterly. Chemical fertilizers restore productivity for the next yield. Organic gardeners call these "dead" fields and scorn the exhausting production cycle. Earthworms have long since been kayoed. But as long as millions of people live in cities and look to the supermarkets for survival, it will continue this way.

One seldom hears of that old villain DDT anymore, which tends to lull the consumer into thinking that dangerous pesticides have gone out. Nothing could be farther from the truth. In the spring of 1973 a small AP story datelined Sacramento told of a cropduster airplane crashing in the Valley. The pilot, who had been contaminated by his cargo of "deadly parathion," was immediately rolled into a nearby irrigation ditch and bathed. He was then rushed to a hospital where his condition was pronounced "fair." During the growing seasons one often sees these small planes, light and graceful as dragonflies, darting across the vegetable fields, spraying their chemical clouds that help to guarantee bumper crops of "unflawed" produce.

More recently an early morning explosion in a new manufacturing facility on the outskirts of San Solace sent a number of employees to the emergency burn center, seriously injured from a "flash vapor fire." They were working on the production of agricultural chemicals. Firefighters put on masks to protect themselves from the toxic fumes. Was this the new "clean" industry the county sought? And how would it affect the spaghetti sauce? Perhaps University scientists even now were developing a new breed of people with resistant livers. Post-carcinogenic Man.

Among the small farmers and orchard growers in our county, however, very few any longer use spray, knowing their customers feel reassured by a bit of scale on the fruit or a worm or two in a vegetable.

On the fringes of San Solace, far from the new country club, lie older structures without which the town could not have prospered—the labor camps of the Chicano field workers, most of which decade after decade remained little changed except as they deteriorate. The sad truth is that the people who have done the most to harvest the crops that fill the cornucopia are the very ones who have benefitted least. Just recently, however, things have begun to improve a little for the victims of an ancient caste system perpetuated by the richest industry of the richest state of the richest nation in the world. In the depression of 1975, poverty spread upward into the middle-class and thus, in a sense, became respectable. One result: after years of stalling, federal legislation was finally enacted that extended unemployment insurance to at least some field workers.

Recent California history with respect to such legislation is interesting. In 1972 the California Legislature had already voted to extend the benefits of unemployment compensation to field workers. But the Reagan administration tacked onto the bill an amendment that would have required workers to waive their rights to sue growers for damages when they were injured in the fields by poison spray. Moreover, it would have required their dependents to waive all rights to damages for workers killed in the fields! Since no law can require a citizen to waive his lawful rights, Governor Reagan on reflection decided the better part of wisdom was simply to veto the entire legislation, which he did.

Cesar Chavez, who had asked, "Is it too much to want a measure of justice for the poorest people in our land?", received more emphatic answers when Jerry Brown succeeded to the governorship. Chavez at the time lay in a San Jose hospital, his back permanently damaged by use of that symbol of "stoop labor," the short-handled hoe. And then, wonder of wonders! almost immediately California by executive order outlawed the required use of this hoe. Governor Brown then convened round-the-clock negotiations with Chavez's United Farm Workers, the Teamsters' Union, and growers. For the first time in history there is now the promise of harvests uninterrupted by labor strife.

So things are looking up a bit for the field workers and their children, even though the work remains grueling, the wages as low as the law allows, and home can still be a cardboard suitcase in the labor contractor's old bus.

In late December I finished some holiday errands and wound up at a retail meat-freezing establishment on the highway. The place was packed with Chicanos. As I waited for a box of neatly wrapped white packages of beef (this was before the great beef shortage) the butcher took care of his food stamp customers from the labor camp. One woman was so grossly obese from her flour-and-beans diet that she looked as if she could not survive much longer. Another was tiny, twisted, shrunken, and mad of eye. They were all exchanging stamps and money for great plastic sacks of tripe, lights, and other viscera.

Mexican-American soul food was just like the soul food of the poor everywhere.

The cheerful Italian butcher gave them all calendars and, to the fat woman, said, "Now don't forget. If you get a bottle before Christmas, come in to see me!"

From there I went to buy ropes of gift garlic from the braided-garlic man. In the filtered sunlight beneath an old, gnarled California pepper tree on an otherwise vacant lot, he grew rows of sweetcorn so tall and even that you knew he used chemical fertilizers. He grew carrots, onions, red chili peppers, and various herbs for the tourist trade. After the garlic harvest, he braided the stems into ropes and sold them from a small stand. Today I could not find him beneath his pepper tree and traced him to a small frame house. His wife came to the door.

"Come around to the drying shed," she said.

She told me that when she had worked in the fields, they had lunched on bread and raw garlic, a munch of garlic and a crunch of bread—a double staff of life.

As we rounded the corner of the house, I found myself enveloped in the sweet, heavy fragrance of flowers and herbs, and in a delirious burst of music. Birds in cages—parrakeets and budgrigars that she raised for the pet stores—were singing with that particular manic joy that birds alone express, even those that have never in their lives known the outside of a cage. I felt like an intruder in this private corner of the Old World.

For hours afterward I carried in my head the remembrance of color, song, and perfume. And for days thereafter, my car still carried the bracing, distinctive aroma of fresh garlic.

Chapter XV

# Freeway & Charter Flight

In morning darkness I found myself on the Nimitz Freeway leaving the Bay Area after a few days of business and pleasure. The Nimitz goes over the top of one side of Oakland which, if you know Oakland, you will realize is the best way to go.

There I was, hitting the early commuter traffic at a bit after six o'clock on the Nimitz, hurrying toward the country, and it was all coming over me again as it had a thousand times. Panic. Terror. Screams. Blood. Freeway phobia. I have a memory for traffic fatalities as some people can recite batting averages. Ask me how many people died in split-second slow-motion crashes on Labor Day Weekend 1945, and Primary Causes. On this particular morning I knew, for example, that it would be only six months until the total U.S. traffic deaths stood at two million—more than all our war deaths in history, far more. Why were the fallen not interred with full honors in a National Cemetery at the outskirts of Detroit?

On the short stretch from Berkeley to Oakland I had kept to a slow lane, feeling calm, confident, able to cope, relishing the pleasant memories of having seen old friends and looking forward to a long life filled with reasonably good works and extraordinary rewards. And then—you must believe it—in the murderous twinkle of a freeway engineer's eye the slow lane, in funneling me onto the Nimitz, has become the next-to-fastest lane. Suddenly the yellow eyes of all the of hell are closing in, in twenty-nine lanes, bearing down at high speed while ahead, a sea of red taillights pulls and mesmerizes me. We are entering the long curve now where defective construction causes one's car to bounce like a small boat crossing the North Atlantic on the bias.

My palms are slick. I want to move over into the third lane and on to the fourth and fifth until I reach the twenty-ninth or slowest lane, but I dare not for the closing pack will crash into me and then ricochet into all the others. This time it is too much. I will lose control. I will throw up my hands and scream. Then the world will pull down black-out curtains, except for the shocking-pink flickering of the emergency traffic flares.

We shall all be lying there on the cold concrete with our shoes off. Traffic accidents always knock off one's shoes. The pilot of the TV helicopter will report the pall of smoke, the burning rubber and hot metal; and we shall all be dead.

First, however, I shall observe the dashboard crumpling slowly toward me. I will think calmly, slowly, "Now we are all dead and it is a great shame. I was not ready to die. This is most unfair." Ambulances and police sirens are moaning. The driver of a car that was not involved sees us without our shoes among the seared rubber and the burning metal. He suffers a heart attack and is rushed off in an ambulance. Shoeless into eternity.

So ended only one of my freeway nightmares. I made my selection and played it as I drove, the way people with steelier nerves might choose a tape from a cassette. But their choice was voluntary. Mine was inescapable.

How many millions of people hated the freeways as I did? We were all triply deprived by them. First, by the alternative and always more interesting ways of travelling between cities, the back roads that had been gobbled up as a python might consume earthworms; or travelling by a comfortable train with good schedules between metropolitan areas and the country. Thirdly we were deprived of the native grass, trees, and blossoms replaced at great cost of our tax revenues by concrete and dusty oleanders and other uniform growth.

Now the country hills lay ahead and I recovered my cool. When at last I turned off the freeway and drove toward San Solace and on into the canyon, the car moved almost soundlessly between drifted rows of winter leaves. At the swamp I drove through the gate and up the drive to the house where the dog leaped at the wire, a dog resembling a gold ball in cold-weather coat.

At that moment I knew this small place squeegeed out of a marsh had acquired the certain identity of home, the place I could hardly wait to get back to.

But almost at once the telephone rang. In response to a wire I turned right around to drive back toward the Oakland Airport. The charter flight from London was "due" in two hours.

# Chapter XVI

# Enter Clarence

Next day the clothesline flapped triumphantly with purple pantaloons from Koochistan, their drawstring broken, a frayed red jersey stencilled *Ibiza*, many pairs of jeans held together by faded threads, a flamboyant cape from Morocco, and smaller articles of clothing all so worn and faded as to defy classification. All garishly out of place on the swamp. But Alice was happy and so was I. The sky was blue, the day perfect.

Robert arrived from Los Angeles. To celebrate Amanda's return, we rented a chainsaw and took turns slicing up the partially rotted willow tree that had fallen across the creek.

Leaving the tree in long sections for a dam—wrong wood, although I was not so wise then—we climbed the hill and tackled the larger branches of an old oak that was dying. These we cut into lengths for the fireplace. Later I would split up the thicker chunks with a wedge.

In the suburbs of the past I had nursed uncharitable thoughts of any neighbor who shattered my weekend peace with the whine of power saws. Now, having learned this new skill, I was wreaking revenge upon the countryside and finding it sweet. Ah! the simple joy of sending forth wave on wave of pure, eardrum-rupturing noise, polluting the sunlight with clouds of thick, nasty smoke. The chunks of oak fell off like cubes of butter. Imagine a chainsaw being so easy!

Robert, when not otherwise occupied, tends to stand and dream with the inner edges of his shoes pressed together at an angle, his body centered in his personal, powerfully radiant energy field. His revery is filled with UFOs and chemistry and physics and ideas for such things as controlling the temperature of houses by flashing tiny jolts of electricity through glass windowpanes.

After we had finished demolishing weak trees, he went to his van and returned carrying templates for a way of building inexpensive shelters out of small, tough triangles of a lightweight material—without the use of frame, bolts, rivets, or adhesives.

"Earthquake-proof," he assured us, "and fire-resistant."

A week or so after his visit I read with auntish horror an item on the front page of *The Wall Street Journal* announcing that he would demonstrate the fireproof qualities of his dome on a local college campus by drenching a model in gasoline and setting fire to it—with himself inside. (He did. The dome caught fire. TV cameras made home entertainment of it. He emerged unscathed, except in reputation.)

We rested from our labors on the hillside. The intrepid inventor and the intrepid world traveller, who were more like brother and sister than cousins, began talking idly of the pressures of conformity in America.

"Nothing *bores* me so much as having people tell me I ought to get *married*."

"Most married people are so *miserable* that they can't bear to see anyone single and *enjoying* it. I get it all the time too."

"What *I've* started doing is carrying this little newspaper clipping around in my pocket. It gives all the worst statistics they have been able to put together on the high rate of divorce."

Evil chuckles.

"Far out. Most of my girl friends' mothers keep after them to get married. Especially to doctors."

I, the listener, repressed a shudder.

"Dentists are pretty high too. And engineers."

She shot a glance in my direction and added, "Especially freeway engineers."

"Don't turn to me with your complaints," I said. "If you want to change things, work *inside the system*. Study medicine yourself. Marry some nice male nurse who will pay your tuition. Then, when you are a successful brain surgeon and your husband has become a miserable nag of forty with secret drinking problems—*then* get a divorce. And remarry to a younger, more attractive male nurse who is approximately the age of your own son Edgarde—whom I quite forgot to mention. It is all in how you play your cards, how you handle your swabs and scissors. Anyone who is willing to work can get to the top these days."

"I think I'd rather be your daughter the freeway engineer," she said.

"I don't even know anybody who has a doctor," Robert spoke up dreamily and somewhat irrelevantly, "let alone medical insurance."

"If I have to see a doctor or a dentist," Amanda said, "I'll take another charter flight back to where they aren't such a rip-off. You can fly there for the cost of an office call here."

"All right, if you don't care about *sanitation*. I have never heard such unwholesome and seditious talk."

The day was too fine for staying in. In early afternoon we climbed a perpendicular hillside where winter colors glowed. The russet of rotting leaves, delicate ferns, bright-green moss, the yellow and gray of lichen, all against the dark hillside loam that crumbled beneath our shoes. We panted upward to where the trees became red-trunked madrones with glossy, dark-green leaves resembling those of magnolias; and native oak and a few toyons, which are the wild holly of California. At this season they were covered with clusters of red berries.

We came on two little wigwams made by heaping leaves over a structure of branches. At the doorway of each wigwam, the occupants had neatly deposited their droppings, which looked like those of rabbits. Too vulnerable, these huts, for rabbits. Skunks or oppossums? We carefully skirted the little homes without touching them. Higher up the hill we came to a tree so festooned with Spanish moss that it resembled a witch with a wild, six-foot fall of gray-green hair, and here we could not resist standing under it in turn for color snapshots.

Do anything, talk of anything, keep busy. On that beautiful afternoon the thing uppermost in all our minds remained the war, the never-ending unspeakable war.

And therefore, as soon as we had returned, I suggested another activity—that Robert accompany me on my first visit to the feedstore. This was prior to my ordering the Banties from Clarence.

"I have been shy about going there the first time," I said. "I won't know what to ask for."

"Chickens. Everything."

The two of us drove to San Solace.

As I quickly discovered, not only did the feedstore fulfill one's high expectations of adventure but it made things easy for the amateur.

"Look," I said. "All you have to do is add "ena" to the names of animals to get the right feed for them. Goatena, Sheepena, Pigena, Cowena."

"What," asked Robert, "what kind of animal is a start? Here's a sack of something called Startena."

"I don't know, but watch this. I'm going to try them on twenty-five pounds of Henena."

"But you have no hens, Auntie."

He pointed to the sign on the empty, cold chicken incubator: "Baby chicks, 25 cents ea., April."

I went back to inventorying the shelves. Even more than Will's tiny shop, the feedstore was a visual feast. Sacks, bins, bottles, racks, barrels, all manner of goods and grains. Early tubers and seeds, harness, stiffly waxed lariats, rock salt, tin bathtubs, farm tools, ingenious watering devices and feed dispensers, batteries and signs for electrical fences, rolls of fencing, and remedies in bottle and spray can for all sorts of rural ills, animal or vegetable. The latter had ominous names: TRICE KILLS MICE, DEADLY DEWDROP, RATS DROP DEAD! and our old friend 2-4-6-D, a lethal spray guaranteed to stop nasal congestion forever. There were shelves full of bottles labelled BOUNCE-BACK and BLOAT-GUARD, SWEETLIX and MOUS-PRUFE.

It was here that I later made the important, lasting acquaintance of BAG BALM. Every farm eventually hits on some all-purpose remedy. On the farm of my childhood, my mother had treated all of us for most afflictions with SAVE THE HORSE, a foul-smelling liniment reputed to have been made in India of crushed bettles, and which my father bought for his pacers. BAG BALM comes in a square green tin with the sweetfaced likeness of a cow wreathed in clover on the lid and a fulsome picture of her udder on the side. But it smells pleasant and has excellent healing properties. My neighbors used it on everything from rabbits to steers. I mixed it with kerosene and dipped the Banties' sore, encrusted feet in it; and afterward noted with pleasure that all the tiny scratches and cuts on my hands had been healed.

The floor of the feedstore was also a friendly clutter—of heaps and barrels and bags. Steer manure, chemical fertilizers, redwood bark, pet feed, wildbird seed, fresh eggs, a rack of manufacturers' leaflets to help people like me find our way. SILEAGE FACTS, MASH NEWS, GRIT MANAGEMENT, STX-1040. Just name it—.

Clarence came forward—lank, laconic, and to one of my generation vaguely resembling Joel McCrea. He could have been rehearsing for the lead in a TV show called "Feedstore." There were little creases at the corners of his lips from saying "Yep" all the time.

"Yep?" he greeted me.

Determined to carry off the confrontation with some vestige of authority, I straightened my shoulders and blurted, "Twenty pounds of winter rye seed."

Robert, raising his eyebrows briefly, wandered off toward a barrel of waxed lariat rope. Clasping his hands behind his back, he pressed the inner edges of his shoe soles together and let his eyes go into dreamy orbit. He was in a space vehicle making its third pass at the back of the moon.

As for Clarence, I noted with alarm that he had turned a ghostly pallor and wrongly assumed that I had used improper terminology.

"Maybe what I need," I faltered, "is Ryena. I want to disk the rye sprouts under for green manure."

Clarence turned wordlessly and went into the grain storage section. In a moment he returned gamely carrying a gunny sack of seed. He set it down near a floor scale. He found a tin measuring tray and a paper bag and knelt down beside the grain.

"Boy, do I hate this," he said.

"What's the matter?"

"Hay fever. Sometimes I get it so bad I'm in bed for a week. It really knocks me out."

"Would you like me to do that?"

"I shore would appreciate it."

I squatted by the scales and measured out twenty pounds of rye seed.

"Looks like you're in the wrong job," I said inspiredly.

"Yep."

"If it's an occupational illness, you could probably get full disability compensation."

"Yep."

"Or sue them, the capitalist pigs. They don't care what they do to workers."

"I'm the owner," he said.

"It's still dog-eat-dog, Clarence."

"Yep."

He carried the rye seed to my car, saying "Boy, I shore do appreciate that."

I returned to look for Robert. The store had suddenly filled up with farmers in Stetsons and cowboy boots, who reeked of sheep-dip. They seemed to be ordering enough rock salt to tide their rocks over into the next century. Clarence was beginning to look harried as well as sick. I collected Robert over by the horse beans and white onion sets where he was still in full orbit. We walked through the granary to the hay barn.

The former was a grand, gloomy, high-ceilinged warehouse resembling a medieval castle. The grain silos were built of two-by-four timbers stacked bricklike in towers that reached to the dusty skylights—presumably to repel rodents with their thickness. The air in here was sweet-smelling, a blend of dust and many grains.

A boy named Bob came forward and I brashly ordered a bale of straw for the future Banty chicks. He led us out onto a loading platform near a pair of new aluminum farm gates marked *SOLD*. A bowlegged, leather-faced man staggered past us at a run, carrying a hundred-pound sack of Scratch on one shoulder and a bale of alfalfa on the other. I could not help feeling a stab of envy. These things proclaimed that he had animals on his farm.

"No steps down," Bob warned, vaulting four feet to the ground. He did not look back. I jumped and Robert followed. It felt good to jump four feet. We reached the barn and Bob loaded the bale into the trunk of my compact.

Driving home, I said, "That's a great place."

"What?" Robert said. "Where?"

After that, whenever I went to town, I made excuses to go to the feed-store. The excuses, as time went on, got increasingly expensive. But Clarence and I got well acquainted.

"Promise me," I said, "that you'll phone me the *minute* you get your Banty chicks in."

Clarence sneezed.

# Chapter XVII

# A Definition of Freedom

On an evening in early spring I looked out the kitchen door to see an enormous owl gliding into the cottonwood tree nearest the house. I whipped into the bedroom for binoculars and back to the door. The sky was almost dark yet I could plainly see the white markings below the face, the rather small round eyes, the gray-speckled front, and the tall ears leaning slightly inward. Closing the door softly, I ran to the bookshelf, found the book, found its picture. "The Great Horned owl," I read, "is common in California."

Almost any bird described as common was now either scarce or rare. It was a thrill to see so closely the friend whose voice often hooted me to sleep. I returned to the door again. The owl, apparently unfrightened by my comings and goings, cried, "Whoo—whoowhoo—whooo!" I, unable to resist, uttered a low hoot. The owl looked at me. Then it looked at Alice who was also looking at me. The owl's dark profile turned away and he paid no further attention to us. I had the uneasy feeling that *we* had been classified as "common in California."

The nicest sound at night was when the male and female on opposite sides of the canyon were calling to each other, exchanging news and hunting tips as they moved through the woods. Their soft, husky calls would be close, then farther away, again nearby, the female's hooting slightly higher than the male's. Then both would move farther and farther off until finally sleep came to me. Sometimes just at dawn I would hear them again, hooting excitedly over the night's catch before they too sought their beds, and reminding me a little of women who had just returned from a sale with good bargains.

The dying oak across the creek was one of the places where they ate and dropped small rodent bones. Another was near an elderberry tree on the county road where I had often seen an owl settling just at dusk and where Alice found the regurgitated remains of a gopher. It was an oblong clot of jellied fur with the two long rodent's incisors still in position. Looking at it, I could not help wincing. Owls are said to be such voracious hunters that they are forced to change their breeding ground every three or four years. I felt reassured. From what I had seen of the gophers of Meanwhile, the owls could live here for centuries.

Perhaps the most disturbing thing about having returned to the country was what ought to have been the easiest—the simple, direct acceptance of beauty. Scales of remembered ugliness, the constraints of habit, and an inescapable awareness interposed themselves—the "knowledge" as Josephine Johnson writes in *The Inland Island*, "that all over the world human beings wake in prisons, wake in hospitals, wake in pain. Wake, and their pain does not pass."

One sought to escape for a while but sooner or later the ghosts would intrude between one's eyes and the most perfect morning spread out frosty, bright, and inviting. I would look down the meadow toward the western hills, hearing the early, drowsy twitterings, feeling that I needed to expand physically to accept so much beauty.

At such times it seemed likely to me that most of us did little for each other except impose greater burdens than the impersonal world contrived for us and that, for ourselves in this hardening process, we built invisible barricades against the healing that nature offered abundantly. I also thought, however, that were I really able to screen out all past conditioning of my ways of seeing, able to erase this greasy film, this interference, this bad life, the shock of perception would be too great. I thought a person could die from it. Was there any other risk worth taking?

On stormy nights when wind grappled at the new roof, I planned my moves for spring and reread old rural favorites.

One thing I meant to do when the weather warmed up was find out what lay under other things outdoors. On a cold day I had casually spaded up a huge clump of coarse grass, only to find that its sod was the winter home of hundreds of dormant red ladybugs. Some of them became embedded in the mud on my shovel and I suppose they eventually froze as a result of my thoughtless and unwarranted intrusion. A dead ladybug was no friend.

Rereading Thoreau, I found his modesty unworthy. Thinking to excuse the Sybaritic excesses of his life at Walden, he wrote: "I should not talk so much about myself if there were anyone else whom I knew as well."

More to my taste was Buckminster Fuller's comment: "The level of literacy about me is constantly rising."

For my part, I should not dream of writing so much about myself were there anyone else who interested me a tenth as much. And that's the truth.

Thoreau would have disapproved of my plans for the swamp. Too much work at the expense of true wealth, the "proper business" of the mind. Don't worry, H. D., it's mostly talk. I plan a very small orchard, knowing the deer and gophers will consume perhaps one-half. As to the vineyard—of course I would plant the grapes. But making one's own wine tended to be outrageously expensive and if the home vintner would admit it—as I did—one would find it hard to top the excellent small wineries in this area. Perhaps—sometimes meanness appeals to me for its own sake—I might buy cheap wine in a liquor store, soak off the labels and glue on my own, rolling such beautiful names across my tongue as Meanwhile Pinot Chardonnay, Meanwhile Petite Sirah, and watching my guests' discomfiture as they felt compelled to utter praise.

The independence of the Thoreauvian mind is today more pertinent than ever. My favorite anecdote concerns his going to his female tailor to order a new suit, having in mind a clear idea of what he wanted— such as keeping certain little touches that the stylists had decreed passé.

Each time his tailoress would tell him gravely, " 'They do not make them so now,' not emphasizing the 'They' at all, as if she quoted an authority as impersonal as the Fates, and I find it difficult to get made what I want, simply because she cannot believe that I mean what I say, that I am so rash. When I hear this oracular sentence, I am for a moment absorbed in thought, emphasizing to myself each word separately that I may come at the meaning of it, that I may find out by what degree of consanguinity *They* are related to *me*, and what authority they may have in an affair which affects me so nearly; and, finally, I am inclined to answer her with equal mystery, and without any more emphasis of the "they"—"It is true, they did not make them so recently, but they do now.' "*

And he noted, "The head monkey at Paris puts on a traveller's cap, and all the monkeys in America do the same. I sometimes despair of getting anything quite simple and honest done in this world by the help of man."

The Countess Lillian felt much the same way which of course was why she had seen fit to learn a variety of skills. And like Thoreau, she wore some of her well-made English clothing—tweeds, silk, and

---

*\*Walden*, H. D. Thoreau, Modern Library, Random House, Inc., New York, 1937.

bombazine—for however long the fabric held together. It was not a case of clothes making the woman but simply of putting inanimate material in its proper relationship.

I called on the Countess one day at her San Francisco house, arriving just as an electrician was leaving with his tray of tools. This surprised me greatly. I found a note from her asking me to wait.

The electrician had been replacing ceiling lights. As he prepared to leave I noticed a bare wall-switch in the front hallway, the metal plate for which lay on the floor.

"Aren't you going to put that on?" I asked, since it seemed obviously an oversight.

"Nope," he replied. "The Countess didn't say to. O' course if she wants it done, I'll do it. But I'll have to get some other tools and come back tomorrow. See, this needs to be shimmed out because of these new wall panels. Can't do it with these tools. Takes a special set."

Shortly afterward the Countess herself returned, carrying a few purchases, including a nice ear trumpet she had found at the Good Will and which she felt might be useful for her increasing involvement in civic affairs. I faithfully reported why the wall-plate still lay on the floor.

She set her parcels down.

"Do you have a hairpin on your person? Well, never mind, this nail file will do."

In a jiffy she had pried forward a metal frame surrounding the light switch. With her other hand she picked up the wall-plate from the floor and snapped it neatly into place.

"*There* we are. I suppose I should join a different guild for this speciality. I hope you won't rat on me."

Later over tea she expanded on her personal philosophy.

"One definition of freedom is not having to wait around for people who are more incompetent than one's self."

"Yes but—," I said. "But you are wealthy and have time to master every skill in the book. Which isn't fair because you can also afford to call in a skilled incompetent."

"Time is for trading. It isn't that we have a great deal of leisure time in post-industrial society, but that we do have many choices and options to trade it for. Although Veblen's theory of conspicuous consumption was developed in the last century, most Americans go right on altruistically supporting manufacturers, middlemen, and their indolent heirs as if they had no other choices for the use of their money/time. You Yanks baffle me."

"That reminds me," I said, "I have just had a stroke of luck in finding fence builders for the farm—a job bigger than I care to tackle, so I am planning to trade money for it."

I told her about Tom and Marcie Oleander and their communard Dave, who were caretaking Dr. Bufano's Cattle & Vegetable Farm up the canyon road behind the sign with the big yellow daisy.

"It was funny the way I found them—or they, me," I said. "Mysterious forces are at work in these country towns. They are shaping our destinies while we sleep."

Chapter XVIII

# The Depression-Makers

Tom and Dave stood on the porch and introduced themselves while behind them, their old green pickup truck pinged, heaved an enormous sigh, and settled on its worn tires.

Tom was a lanky Oklahoman in his midtwenties who tied his long hair back with a rubberband. Dave, who looked younger and affected the role of silent partner, turned out to be shy but noticing as well as fond of owls, which of course I found appealing. Tom's bib overalls were womansewn with a jaunty red trim at the bottoms where they grounded on size 11 Army Surplus boots. Dave, possibly because he and his old lady had recently split up, lacked adornment.

"When I stopped at the hardware store this morning," Tom drawled, "Mr. Hotmeier told me some woman had been in, asking about fence builders. He didn't remember the name or the phone number or where she lived—except he thought she lived in the canyon."

"How did you happen to find exactly the right place?" I asked, inviting them in.

"Dunno," Tom said, scratching his head and flashing a smile that lighted his grass-green eyes. "The Lord moves in mysterious ways. Mr. Hotmeier went to the telephone. And when he came back he said it was probably you."

"But I haven't told a single other person about needing fence builders."

"Like I say," Tom said. "The Lord—."

"Far out," Dave contributed.

"Well," I said, dismissing the supernatural, "How are you at building things. Fences, a small bridge, a gate?"

"We can build just about anything," Tom said. "Built the cabin on the ranch we're caretaking, and all the fences. Cleared the land with hand tools, because like we figgered to grow organic vegetables and start our own route. And we used the oak trees that had to be cleared to make posts for the fences."

"How long have you lived there?"

"About a year."

"Sounds as if you've done a lot."

"Nah. Probably the most important thing is my wife Marcie and I, we've managed to get it together—that is, deciding on how we want to spend our lives. That was the hardest part."

"Please sit down," I said. "How *do* you plan to spend your lives?"

"We're going to make it outside the system, that's for sure," Tom drawled. "We want to do it working with other people—not necessarily in a commune but out on the land with people who have the same goals. I see it as doing two things: I'm pulling my energy out of the system in hopes it will collapse all the sooner and I'm putting it into finding a better way to live."

He assured me that the old institutions were done for, no question about it, adding "We're learning to be as self-sufficient as we can."

Crossing the haul road I indicated trees freshly broken off and mashed back by the road maintenance bulldozer of the Benevolent Cement Co.

"If you've read the glowing accounts in their company newsletter about what great conservationists they are, you would find this hard to believe."

"The truth never comes until you're ready for it," Tom said with the ring of a Mormon Elder. "It's the same with these people now—."

"When will you come to work?" I asked.

"We'd like to start right away," Tom said, "because we've had a kind of altercation with Dr. Bufano and we have to move pretty soon. He doesn't see things the way we do. And we feel he has gone back on our understanding. So we want to look for a new piece of land as soon as we can."

"What was the trouble?" I asked, on my usual premise that when a thing is none of your business, asking is the only way you are likely to find out.

"Well, when we took over the 'leven hundred acres, we were to do certain things in exchange for living there and growing vegetables. As an example of the problems—next week he's sending someone up to the place to spray the thistles. Now, we just don't see how we can sell organic vegetables in good conscience after spray has been used."

Undeveloped range land in California, incidentally, is often owned by absentee doctors or lawyers such as the Oleanders' patron, who acquire it for tax write-off purposes. Although the professional might solve his tax problems more directly by cutting his fees and helping his patients and clients, or by reducing his working hours and enjoying more leisure, this was seldom done.

"What will you do now?" I asked.

"We're going to build a camper for the truck and figger to be moving on in about a month," said Tom. "Can't wait too long because the baby's due pretty soon. We'd like to find a place even farther from towns than we are now. We can smell the Cement Company's smoke 'way up there in the hills. I don't know where we'll go but I know we'll find just the place we're lookin' for."

They came to work the next day. We were having a spell of unseasonally hot weather for early spring.

I was sitting on the grass that afternoon, looking at illustrations in a book on astronomy when they came crashing down from the hillside, redfaced and panting, to refill their water jug. Tom peered over my shoulder at the diagrams of the constellations and, referring I guess to Copernicus, said, "Wow! That guy must have been *really* spaced out!"

They flopped down and seemed to be laughing at some private joke. Since stretching barbed wire through that brush on a hot day was grueling work, I asked why they were amused.

"Dave threw the coins last night," Tom explained. "*The I Ching* predicted 'exhaustion.'"

They started talking about what they thought was happening to the country.

"It's like we figure there's bound to be a second Great Depression," Tom said. "Partly because of so many young people like us who have dropped out of the system. I was in the Marines in 'Nam, which was when I saw what bad shape our country was in. I decided that things couldn't go on like that, that we just had to change, and what it came down to was each person taking the responsibility."

I asked Dave if he felt the same way.

"For sure," he said. "I've travelled around—lived on two or three communes."

"Like it?"

"It was hard for me to adjust to not having privacy, living so close with people, but it was probably the most important thing I've ever done."

"All owl-lovers need quite a bit of solitude," I said.

"I've done all kinds of crazy things," Dave continued. "Acted like a bum sometimes; or other times, making a lot of money and spending it all at once. Being hungry—I've done that. But that was never the worst part. That was just nothing. It's amazing how easily you can get the few things you need, but you have to look for them. It's the looking that's the best part. And the funny way things turn up sometimes, just when you need them.

"Like all across the country now, when you meet other people, you can tell from looking at them, 'They're on my level,' and it's like a big neighborhood, or a family, because everybody helps everybody else. No, it's not hard once you get into it."

"When we move on from here," Tom added, "we figure to try to find maybe three or four people who would just like to go out into the woods and try to make it together. And like Dave says, it's real easy. We could probably buy some land if we wanted to just live in our camper and work at jobs for a while. The hardest thing for us, because we live so easy-like, is decidin' where we want to go next. There are so many options. Makes it hard to focus, to narrow things down."

I said, "It's odd. The greatest difference between your generation and mine is that mine went through the Depression. So, out of the need to survive, we grew up fearful and everlastingly in need. Even after we had social security, fringe benefits, and all. As the middle class bulged, corruption became ever more respectable. But now *you* plan to create a new depression in reaction to what you inherited from our depression, which is just the corruption you are all grizzling about. I must say, I am puzzled."

Tom grinned, shuffled his No. 11's, and said, "I reckon we've still got to work out some of the details. Right now, my main goal is to find an old lady for Dave. And then o'course with the baby almost due, we've got to lay back and find a better spot."

"Your first baby?"

The grin broadened.

"Yep. We figger to raise a bunch of little illiterate Okies here in California."

Tom said he did not want his kids forced to attend public schools.

"We don't figger to register our baby's birth. I don't want my kids growing up just to be slaves for the government. We realize, my wife and I, that if our kids don't go to regular schools, we'll have to offer them an alternative way so that they can cope with life."

Marcie, he said, was studying deep-breathing control so that she would know what to do when he delivered the baby, in their camper or wherever they happened to be. Her mother, as Marcie later told me, had been deprived of the full joys of childbirth by being hospitalized, anaesthetized, and even having her breasts bound. Now it seemed that her daughter could only experience joy through pain. As with poverty, they seemed to need and crave it to sharpen the sense of life.

Tom the tribal patriarch was quick to protect his little group from any revisionist nonsense or misguided charity on my part. Dave was denied the borrowing of my current periodicals, even though his eyes brightened at my offer. Tom said that what with takin' turns readin' the Trilogy in the cabin at night, it didn't leave them a whole lot o' time. Especially when you had been layin' back, buildin' fence all day.

Then when I asked Marcie later if they wanted some left-over rice that Amanda had gotten through the "food conspiracy" or bulk-buying project at her school, she quickly said, "No, we don't need it." And even when I said it was full of husks, she politely stood firm, making me feel like a tarnished pusher.

I was, however, delighted by their independence, being bored with the martyred elders of my own generation. Too often I had heard the latter shrilling, "They'll wind up on *welfare* and *we'll* have to *support* them." And of course the martyrs lived for this day. To see one's termite-riddled institutions scored was bad enough, but to have one's niggling charity rejected—that was the living-bloody-end.

But I also felt sad, knowing that idealists with staying power were extremely rare in any generation.

Baiting Tom a little, I said, "I've noticed that most young people are still going along with the system. At its best, you have to admit, it can be seductive. School, job, promotions—and before you know it, tunnel vision just like their parents'. Because making it, for most people, offers their only chance to use their best skills."

Tom said, sensibly, "It's the bread. The people we know, they may want to move to the land but they're not ready to just start out the way we did and try to make it. So they think—we'll work a while and earn some money; pay off our debts; and just as soon as we've got enough money we'll split. But even the ones who stay with the system, or who go into businesses for themselves, are going to do things differently. This country just has to change."

He flashed his patriarchal smile, half apologetic.

"Sound like a dime-store philosopher."

Dave said, "For sure."

They built good, solid fences at Meanwhile, and afterward a redwood bridge with rope handholds across the steep creek gully.

"We're chargin' less for the bridge," Tom explained, "because we enjoyed doing that."

They had finished building their camper and were leaving that day for good. I handed Tom a check. He turned it over in his big hands and murmurred something about a check being "a very abstract thing."

He thanked me and shook me up a little by saying I was "an instrument of the Lord."

He explained that I had been delivered to them just at the moment when they needed bread for moving on.

"I want to see your camper," I said, to cover my discomfort in the role, and we went outdoors.

They displayed it proudly, the tiny, carefully wrought home on top of the old pickup. Dave had painted a plaque for the rear of it, to disarm the hostile elements, on which he had inscribed the lyrics later quoted to me in Marcie's letter. Beneath the roof of that rustic crèche, Tom would deliver the baby, unless they found a new home in time to build more substantial shelter.

I bade them Godspeed and good bread down the road.

Fifteen minutes later the telephone rang.

A woman's voice at my bank asked if it was all right to cash the check I had given them.

"Sure," I said. "It's just a piece of paper."

# Chapter XIX

# Chicken Psych. 1-A

Clarence called from the feedstore and I flew right over.

He handed me a tiny, warm carton with holes punched in the top, through which emerged a frenzy of peeps and cheeps.

"Doesn't look like much of a cash crop, does it? How many?"

"Ten," Clarence said. "Don't forget your Startena."

"Promise to save me any more that come in."

"Yep."

Neither he nor any other familiars of Banties troubled to enlighten me that ten Banties could in a very short time become ten times ten by the simple expedient of flying over, under, and through the fence and hiding their nests. Some chickens, it is true, are so dumb that they cannot see an open gate. Banties are less like chickens than like quail or other wild birds. Under those little topknots they are scheming, scheming. And just about the only thing they are scheming about is Motherhood.

In the event that insects outlast the human species as the research of some scientists indicates may be the case, I wager there will be Banties around to devour the bugs. Banties carry the maternal instinct to preposterous levels. People who raise swans, peacocks, and exotic geese and ducks tend to look down their noses at your mixed Banty poultry flock. The unsettling truth of the matter, however, is that were it not for the superior nesting habits of the Banty hen, there would be very few of these larger fowl produced outside of incubators.

Who ever sees a peahen raising her own young? The minute they get into a spot of trouble, everyone panics and they send for the old Banty midwife. She arrives in culottes on her bicycle, carrying warm flannels and elderberry tea, singing her Baptist hymns in a shrill but steadying way, and takes over.

Or—this is more likely—she was there from the start. Farmers who raise exotic fowl are forever slipping gross eggs under the broody Banty. Few farmers think to consult the Banty as to whether she desires to nurture this mixed bag of monsters. And once she has done so, too few are willing to give her credit.

Later on even I played an unfair trick on the Banties, buying a dozen fertile brown eggs from a health food store and slipping them under the broodies.

"What kind of eggs are they?" I asked.

"The farmer said, 'All *I* can say is, they're supposed to be from chickens,' " the clerk told me. She smiled reassuringly.

The results, as it turned out, lent more distinction to Meanwhile than we needed—especially among the type of farmer who feels strongly about genetic purity.

But that first day when Alice and I got home with the Banty chicks, all this was a new and thrilling business. Not only were we unaware of the maternal addiction of Banty hens but we little guessed the role this tiny brood had in store for us.

I transferred the Meanwhile Ten into a cage on top of a kitchen counter. They were fuzzy and so tiny I expected to see bits of shell still clinging to them, and very bright of eye. They all knew exactly what to do about survival from the moment they emerged from the box. I could not keep my eyes off them. Alice could not stop salivating.

I placed a lidful of water in the cage. A tiny chick sampled it, tilted back its head, and swallowed in a most graceful gesture. Instantly the nine other chicks raced over and began swallowing with identical beauty of movement. I should have been tipped off then, I suppose, by their natural miming. I filled a tin piepan with mash and set it into the cage. The Banties gobbled up every crumb and then, as one, they all piled into the empty tin and fell asleep with their heads on each other's fuzzy shoulders, first carefully remembering to close their eyelids from the bottoms upward. The pan was now a fuzzy, palpating Banty pie.

Alice, who was standing with her paws on the counter edge and her nose creased against the wire, gave me a sheepish look.

Some of the chicks were yellow with three black lines down their backs like baby pheasants. Some were brown with black markings. Most had long fuzz on their legs and feet that would turn into distinctive feathered leggings. The roosters would have the most elegant anklets and the largest combs and the showiest tailfeathers. The little hens would have cute speckly topknots. They would all have the same bright amber eyes they were born with.

By the time the panful of Banties woke up, Alice and I both had cramps in our legs. Her nose had an X on the tip of it, her lips were sucked up in crinkles along the edges of her teeth, and her cheeks kept whiffling in and out with emotional tension. As I have mentioned, she is a retriever.

"Down, Alice," I said.

She failed to hear. I repeated the command twice more. Then a shudder swept through her body, ending in a long, tortured sigh, and she dropped to the floor.

"Alice," I said, "*Chick-chicks* are a *no-no!*"

She groaned and lay down.

A fly zoomed through the cage. One Banty dived and caught it. This probably was the first time any of them had ever seen a fly but the catcher instantly knew it had a prize and was supposed to run with it. To the others, it was immediately clear that he must be chased. For a moment the entire field was in violent motion.

Forcing myself, I left them and turned to my work.

That night I placed an electric lamp on top of the cage, and draped a towel around the cage to keep out drafts. When the lamp was on, the chicks always arranged themselves in a tight little pile directly beneath it. But as soon as they began to sprout a few wing-feathers, they spread out and tried different sleeping formations, such as lying in two straight lines with the heads of the rear chicks on the backs of the front-liners. I worried about the heads of those in front falling through the wire and being snapped off by a vicious Farkle. I fretted too about the light bulb burning out in the night, and the Ten dying of cold. On both counts, I could have saved myself the trouble.

When their heat lamp went off, the chicks squawked loudly enough to waken the whole household.

As for Ferd and Fanny Farkle, they inspected the cage, sniffed the caglings, and quickly walked away, shaking their feet as if they had just waded through a barnyard. It was the same gesture they used if they did not care for the food I had put in their dish. They were exhaling with nasty expressions on their faces. Unlike Alice, they understood at once that caged chicks were different from wild birds. (If they were wild birds, as the smart Farkles reasoned, why would they smell like chick-enshit?)

One day, though, when the Banties were loose in the yard, Ferd, seeing that I was dressed for town and therefore was preparing to abandon him, dashed in front of me, scooped up a Banty chick in his mouth, flashed across the yard and into the workshop where he leaped up into a loft. I followed quite slowly, assuming that the damage had been done and not really wishing to see the corpse. But when I got to the workshop, Ferd after a minute or two jumped down from the loft, released the baby chick at my feet, and raced dramatically out into the yard.

The chick, without a toothmark on it, gave a peep and dashed back to its group. I had seldom seen a neater display of Ferdmanship.

Everyday, exactly like human infants, the Ten learned a new trick which they would practice incessantly until they had it down pat, or had learned another trick, whichever came first.

One day they learned the shimmy, which is basic scratching. So far as I know, no choreographer has credited chickens for their contribution to this extremely complicated folk art. It starts out slowly with the chick on one foot, shimmying back and forth, then shifting its balance to the other foot and repeating the procedure. But when the momentum speeds up and becomes smooth and cool and professional, you can scarcely see the feet for flying straw. One chick started it, nine enviously followed suit, and soon the cage was bare of straw and the counter and floor littered with it. Among other things it was obvious that the Ten were getting beyond the kitchen-counter stage.

The shimmy-scratch, incidentally, is one of the first things the mother Banty teaches and under her tutelage it is an arduous introduction to life. The chicks spend half of the first few days flying upside-down through the air, tangled up in weeds, while the mother thinks only of worms. But the Ten knew nothing of mother. They were incubator babes—which bore ominously on the fate in store for Alice and me.

I went to the lumber yard which is a nice place for scrounging because they keep odds and ends of milled stuff. While I was ordering five gallons of Thompson's water seal from Virgil, to brush on the redwood siding of the house and workshop, my eyes were darting around. Finally I spotted an old redwood frame with a screen on it, lying on a heap of waste lumber.

"That would make a pen for the Banties until the chicken house is built," I said.

Virgil's eyes beneath his yellow hardhat brightened.

"You raising Banties out at the farm? You want that?"

He scrambled up and pulled it down.

"Boy!" he continued. "Go into business with me and I'll come up and build you a chicken house. There's just *nothing* as good as fresh Banty eggs."

"I got them mainly to eat bugs in the garden, but it will be good to have eggs. Do they lay very many?"

"No," Virgil said. "A Banty lays an egg about every other day. But wow! I know people around here who would give 'most anything to be able to buy Banty eggs."

"That settles it then—about the cash crop."

"Where's your car? I'll tie this on for you."

While tying, he conducted an enthusiastic testimonial to certain other delicacies that came close to being almost as good as Banty eggs but were not and could not be in a million years.

"The difference between Banty eggs and regular eggs," he said, "is like the difference between Maine lobster and West Coast lobster. Of course, your *best* lobster is the Australian kind. *It's* like the difference between West Coast crab and Alaska king crab, or—here. I better stop tying so many knots in this rope or you'll never be able to get it off.

"What makes the taste difference with Banties is that they're always out eating green grass and protein-rich bugs."

"I'll bring you some as soon as the production line is rolling," I promised.

At home, no sooner had I converted the screen frame into the front of a roomy outdoor cage and emptied the Meanwhile Ten into it than Alice plopped herself down right up against the screen. Sucking up her mouth in a mad, grimacing smile, she lay there hour after hour, refusing to budge. As the chicks raced about, her head whipped back and forth. Friend or fiend, I wondered? Guardian or consumer? In either event, the poor dog's anxiety was terrible to see.

One chick scratched its eyebrow with its toenail and all followed suit. One stretched its legs backward like an exercising ballerina and all performed this graceful gesture. Sometimes they all sat down in their mash pan and practiced dust bathing. (My only excuse for exposing the reader to so *much* information about Banties is, quite simply, because a little learning can be an extremely dangerous thing.) A day came, before long, when these tiny paragons discovered they could fly. The first one did a little up-and-down-and-whirl which, of course, excited the whole group. All ten did whirlies then, right under each others' noses and it turned into miniature fights. Chicken behaviorists claim that at some point in the life of a caged flock, the entire pecking order is for all time established in a single flurry of confrontations. I found, however, that the pecking order of the Meanwhile Ten remained in some flux. Also, contrary to what the experts say, the lowest chicken in the pecking order in a cage will not be pecked to death—unless the cage happens to be overcrowded.

In a very small way, by giving the Banties food in various places and plenty of space, I was able to reverse a miserable and sadistic trend in modern poultry-raising.

Almost any creature that nowadays finds its way onto a supermarket meat or egg display counter has spent its entire brief existence crammed into a tiny cage or feedlot, unable to walk or move, while being incessantly stuffed with hoked-up superfeed, and kept awake with electric lights to squeeze more and more production from it. It is a matter for consumer outrage and boycotting. And it is a very strong argument, among several, to be made for increasing the number of small subsistence farms.

Everyone who raises fowl or livestock on a farm must face the necessity of killing them at times. But there is never any need for cruelty or for any greater natural pain than a human being faces in the course of life.

One morning when I opened the Banties' cage, they all came flying forth like a horde of miniature helicopters. And that was the day Alice and I received our first lesson in Chicken Psych. 1-A. Unbeknownst to us, the Ten had imprinted our likenesses upon their brains as mother figures.

They rushed toward us, uttering joyous shrieks, and fell in behind us, wanting to be led somewhere. It was never possible to shoo these chickens. I could flap my arms and yell until I was hoarse but they simply kept sneaking around and falling in behind their appointed leaders. So *I* got the picture, accepted my role, and started leading them into the chicken house or to the garden where perhaps I had just dug up a nest of inch-worms. But it was rough on poor Alice who simply could not adjust to a pack of chickens nibbling at her heels.

First she killed one of them. Then, in a terrible burst of afternoon carnage, she disposed of four more. That time I was nearby and punished her at once. I feared, however, that she might never be able to conquer her instincts.

For about a month, I shut her up in her kennel whenever I went to let the chickens out into the field. If I forgot to put her in, I would return from the chickenhouse to find her sitting remorsefully at the kennel gate like an addict waiting to be committed. I still let her go with me in the mornings when I went to feed the chickens in their pen, however, a jaunt that was obviously the high point of her day. She raced ahead with tail flying. But the moment she got to the pen she would lie down to forestall over-reaction. Sometimes the Banties flew right past her nose but she never budged. One day when I went to put her into the kennel as usual, she just stood aside and looked at me. I looked at her. "Okay," I said, but in a very skeptical tone.

However, it turned out that she was absolutely right in her self-assessment. In time we replenished the Banty flock. And the Banty flock replenished the Banty flock many times over. Generations of baby chicks tagged around after Alice or flew past her nose or rushed up to her as to a foster aunt, screaming "Pwa-aa-ak!" But never again did she lose her cool. She became such a model Banty guardian that she let the chickens crowd into her dog house on rainy days, where they stayed to lay their eggs, and sometimes Alice would carry an egg gently to my front door and lay it down.

Once I noticed that she seemed to be abnormally cramped at the entrance of her A-frame doghouse. On investigating, I found fifteen Banty and duck eggs.

Alice also performed an extremely valuable service for me by keeping track of where the tricky little hens hid their nests. I had tried to follow, from a distance, the artful dodging of a particular white Banty that I knew was setting in the woods. When she reached some brambles at the base of a big sycamore tree, she cast a quick glance around and simply vanished. I led Alice to the area and said, "Find the *chick-chick*." She went directly to a heap of brambles and fern and just stood there. I looked and saw only brambles and fern. Alice, getting bored, started to wander off toward the creek. When I called her back and repeated the order, she politely returned to the same spot and simply stood, not pointing, because she is not a rude dog. And not poking her nose into the nest, because she knows how Banties feel about their nests which is how *she* would feel if she laid Banty eggs or happened to be asked by a Banty sometime to keep things warm. But there the nest was, as I discovered when I knelt down and looked closer.

It was the same when Alice on command led me to the mother duck's nest hidden in the swampgrass, a case of her just standing and looking for a moment without actually staring.

One morning I was about to arise when I heard her utter a short, unserious growl in her doghouse. The kind of growl she might give if one of her dog visitors were taking a liberty—an I'm-not-really-mad-yet-but-don't-go-too-far, don't-press-it sort of growl.

I got up, went to the kitchen to get coffee started and Farkles fed, and opened the door with the expectation of finding her waiting with her chipped green saucepan in her mouth. But no Alice. I looked out of windows on all sides of the house. Funny thing. No Alice.

I looked toward the doghouse and there was the mother duck, sitting smack in the doorway, gearing up to lay an egg. Alice, who had been quite prepared to wait out the blessed event even if it meant missing breakfast, came scrambling out with that silly expression on her face.

Among the Banty hens was a tiny, speckled one with a gingerbread-and-white icing topknot who never missed the chance to fly onto my shoulder, dig in her toenails, and fill me in on what was "really going on in the so-called Free World." Her darkest secret, aside from a case of Freudian hysteria, seemed to be the fact that she could not stand other Banties. First Out, Last In, as I called her, for good reason.

They were all insatiably curious and sociable, often wandering up to the French doors, pecking on the glass, and pressing their eyes right up against it to have a look into the living room. They sang constantly in shrill, devotional voices, so that I came to think of them as a delegation of Baptists, carrying their hymn books and very likely up-tight about Original Sin.

Out in the chickenhouse they would sometimes raise a nerve-jangling roundelay expressing panic or hysteria in which the whole flock joined, which might last for as long as fifteen minutes. Whatever set them off—a leaf falling on the roof or a Martian landing or the shadow of a sparrow diving for grain—one Banty would raise the alarm with a shrill, bugling cry. Mob infection would spread until all were shrieking.

Alice was at first distressed by this hideous racket. To her it meant that predators were after her charges or, at the very least that they were being attacked by a horse, a bulldozer, or a pickup truck. She would rush out toward the chickenhouse, lunging on her hindlegs, hackles raised, tail whirling in circles, and in her deepest, most terrifying voice, would cry, "Wff-ff!"

"It's okay, Alice," I would say, "it's just *them*."

She would come back, looking humiliated. Identifying so closely with the Banties, it seemed as if their lapses were her fault.

A day of wild excitement came when one of the Original Baptists (the imprinting group) laid her first egg; and of course the others immediately followed suit—producing a succession of perfectly shaped little ovals with very strong shells and orange yolks. The morning devotional became more ecstatic than ever. I would hold one of these Arpian sculptures in the palm of my hand and marvel at how tiny hens could produce them out of a few grains of corn, some blades of grass, a handful of earwigs, and the metabolism of their insubstantial bodies. I worried about the minerals being leached from their teeth and bones. Preparing to break such an egg into a pan seemed to me almost like cannibalism. Like Alice, I over-identified.

# Chapter XX

# The Cash-Flow Dynamic

Amy and Angelo live up the road on a five-acre farm where, among other things, they have begun to raise Aracauna chickens which lay blue-and green-shelled eggs reputed to be low in cholesterol. John and Serena live down the road on a ten-acre place and have a genuine stained-glass window in their small chickenhouse, indicating that they may be secretly planning to inspire their hens to mutations that will corner the Easter egg market. These facts I mention only by way of pointing up the intensity of competition confronting my mixed Banty flock.

John teaches school. Angelo works as an electrician. They have learned farming only in the past few years and put every spare moment of their time into it. Amy and Serena run an organic vegetable shop downtown in the summer and Amy sells eggs and good vegetables to special customers on a route.

Angelo is the catalyst in our little strip of canyon because he owns a red tractor with a disk, harrow, and row-making attachments, in which he takes inordinate pride. On a Saturday morning it is not unusual to hear his tractor clattering down the county road and his voice singing softly, as if he were still asleep, but the tractor knew the way. This means that none of us has to rent those power tillers so often portrayed in the advertisements as being easily operated by a woman of ninety who has lost one arm in the washing machine. I once did rent, for a day, an old-style power tiller that required *two* hands to operate. The only thing was, it quite obviously did not require any feet. Mine almost never touched the ground in the eight hours it took me to cultivate a tiny garden. If you can imagine being a pygmy hanging onto the blades on the inside of an osterizer, you will get the idea. I decided it was a quicker route to a mastectomy than I desired. It was with the greatest pleasure that we welcomed Angelo and his tractor to Meanwhile. (One of these days, of course, we'll get into the Ruth Stout method.)

Those of us who are new and relatively inexperienced have learned how essential cooperation can be. At a certain point, bartering and the cooperative—however informal—begin to substitute for old-style money. The latter buys a lot of things you need, but it doesn't serve as currency for the basic riches.

Knowledge is still the most valuable currency, but it is knowledge of how to fight inflation by augmenting food and energy resources that has come to be our economic real wealth. Among our small new canyon farms, we trade tools, equipment, broody hens, cuttings, and knowledge; and sometimes we order seeds and plants in bulk by mail. When Angelo built his barn, everyone showed up one Sunday, looking rather self-conscious, to help nail on the siding. We also trade farm-sitting services.

Angelo and John raise two steers together each year because between them they have enough grazing land. I and other neighbors buy some of their beef, which we know is lean and has not been shot full of hormones or quick-weight chemicals.

The deer are quite a nuisance at the John and Serena spread, so Angelo and Amy grow at their place the tomatoes, lettuce, and other things that deer desire. John and Serena grow the sweetcorn and perhaps some field corn for the cattle, and anything else obnoxious to deer.

Angelo has a wine-press. I have brashly promised to grow the grapes for it.

Amy has taken over as middlewoman for the Banty eggs of Meanwhile. About once a week she calls with some empty egg cartons. We then get right down to our cash-flow and market manipulations. Even Thurber's fabled Aunt Wilma ("The Figgerin' of Aunt Wilma") would find it no snap to follow these negotiations.

To start with, on the plus side, she has persuaded her customers of the efficacy and wisdom of the Old Chinese Egg Method. Meaning, if you don't wash the mucous coating off the shells, they will keep indefinitely without refrigeration. Not having to wash the eggs saves me a lot of time. And fortunately Amy agrees with my theory that to candle Banty eggs would probably impair their fertility, just as exposure to heat can cause male sterilization, so I don't do that either.

She transfers the eggs into her cartons, takes out her wallet and says, "Let's see, that's four dozen, I owe you $1.75. I have $2.00, do you have a quarter?"

"No," I say, after searching my change purse and the petty cash cup. (I am wondering, but I never ask, how she divides four evenly into

$1.75. It involves the size of the eggs and her tiny comission.) "But I owe you for the Avon non-aerosol concentrated room spray for Ferd's sandbox, which cost $2.92 including tax. So actually I owe you—?"

"$1.17," says Amy.

"I'll check the chickenhouse and see if there are enough eggs today to make up the difference."

I return with another seventy-five cents' worth of eggs.

"That's 42 cents I owe you."

Amy says, "But I still owe you for the postage on the plant order to Massachusetts."

"I don't know how much that is yet. Tell you what I'll do, Amy, I'll write you a check for 42 cents. Then we'll be all square for now."

We settle it that way. The next time she comes for eggs, I say, "I owe Angelo $1.00 for the used chicken feeder and $1.00 for the chainsaw fuel. So you can subtract that from the eggs today, or vice versa."

But that day, before we even get to our egg calculations, the mailman drives up with the plant shipment from Massachusetts and Amy owes me $2.49 for her share as against the $2.00 I owe to Angelo. In short, I have 49 cents coming *before* we get to the eggs.

She pulls my old check for 42 cents out of her wallet and says, "Here, why don't you cancel this because it'll only confuse my bank balance, and I'll give you seven cents besides."

"If that's the way you feel about it, if my check isn't good enough for you—. Now, how do we stand on the eggs?"

She is counting cartons.

"Well, we've got quite a few here today," she says. "We've got five dozen. That's $2.25. Plus seven cents is $2.32. Here's three dollars. Do you happen to have change of sixty-eight cents?"

"I'll see. Well—look, I've got two dimes. If I give you back the check for 42 cents and the two dimes, you'll owe a penny. Right?"

Amy stares at me a moment. She nods, without conviction, and hands me a penny.

"I guess we're all squared away now," she says, "until the next time."

She folds my check into her wallet, takes the eggs, and starts for her car.

But just then Barbara arrives and they fall into negotiations outside. It seems she has picked up two packs of cigarettes in town for Amy, who is now trying to kick the red raspberry leaf habit—it conflicts with the effects of her osterized lettuce bath, the dregs of which (if memory serves me) have proved so effective in curing her rabbits' ear mites. Barbara hands over the cigarettes and the sales tag for them.

When Barbara enters the house she hands me my worn check for 42 cents and says, "Amy gave me this. I wonder if you'd mind just tearing it up so it won't foul up my bank balance?"

"Okay," I say, miffed. "It's just a piece of paper, after all. Checks are very abstract things. What is money? What is truth?"

As I tear up the check—which is quite capable of being circulated through the neighborhood until it disintegrates—I am remembering those now bloated-seeming checks I used to receive in the city, which I so loyally spent to cover my cruder needs and to keep my psyche together: a form of over-compensation not only for what money can't buy in a city but what it can't help you avoid.

# Chapter XXI

# The Gentle Revolution—
# Shoestring Power

In 1972 only 55.7 percent of Americans eligible to vote, including a wave of eighteen-year-olds accorded the privilege for the first time, bothered to cast their ballots in the Presidential election. And that was well before the shocks and disillusion of Watergate. At about the same time, President Nixon assured author Theodore White, "Our economy is so strong it would take a genius to wreck it."

Geniuses were at work, both within the system and among those who chose another path. Record high unemployment, zooming crime rates, lagging productivity, and inflation characterizing the first half of the seventies attested to at least to a genius for socioeconomic catastrophe.

When longhaired youth in the sixties formed living groups on rural communes it was not necessarily because they loved farming. Nor was it merely because they felt frustrated by a culture whose educational institutions, if they even recognized new thinking, were often quick to expel the purveyors of it. The young communards sought peace and the company of their own kind, with a space in which—like the Oleanders—to get their heads together.

As time went on, however, the ranks of urban exiles were swelled by older adults, usually middle-class, less footloose than formerly, often the holders of professional degrees, who were not dropping out of society except in the sense of trying to find a more rational way of fulfilling their lives.

The trend of the previous hundred years had gone into reverse.

In the 1930's farmers fled from the land to the cities for several reasons: smallholders were squeezed out by the big, automated dairy ranches and by agribusiness; the Dust Bowl ruined Midwesterners, blowing them toward California; and in those Depression years few jobs existed in the small towns.

There are three and one-half million square miles of land area in the United States, an average of twelve and one-half acres for every man, woman, and child. A family of five, under an equitable division, would have sixty acres. But 70 percent of our population is pressed into 1 percent of the land area.

Today we flee the cities to escape dirt, noise, stress, danger, loneliness; seeking a sense of human proportion and community. But it is hard for a person to leave the city, leaving a job, making a new start. By encouraging this new population flow as a matter of national policy however, we could undoubtedly assure a good future for our children in the year 2000. Or shall we—a mere quarter century from now—throw up our hands as we did just recently on discovering that oil supplies were running out? Shall we merely curse and search for political scapegoats on learning with shock that there are no longer vegetables in the market, oxygen in the air, or earthworms and bacteria in the soil? That insects are resistant to the most hideous new chemicals our researchers have been able to devise? That the proteins and nitrogen needed by millions of people for survival have been flushed into septic tanks or have polluted water supplies near feedlots, while our homes and our offices are cold each winter? Or shall we try, each one of us, as best we can, to become a "closed-space system"?

This means simply that each person shall become a production and recycling world in miniature, giving back the energy and oxygen we use. Trees, if we grow them, will absorb and recycle the carbon dioxide from our lungs and our cars. They will give us wood for the stove in winter. Plant scientists are in the early stages of developing strains that will produce their own built-in fertilizer—a "closed-space system" just for us, if we can bridge the survival gap until around 2000.

The kinds of small-town jobs that subsistence farmers need are increasing twice as fast today as jobs in the metropolitan regions, the U.S. Department of Agriculture tells us. The Census Bureau reports that many of the nation's largely rural states, especially in the South, have been growing in population faster than the largely urban states.

New kinds of farms are being built—or, more accurately, new rural living styles are evolving. An end result of population-flow reversal is that town and county governments will begin to exercise new clout at the Federal level. More sophisticated politicians arising at the local levels will be less conservative than their predecessors.

Since inequality has always been more a psychological than a physical problem, closely related to the lack of control of personal destiny, we shall find the small farm increasingly important as a therapeutic leveller. It provides a sense of autonomy and independence for dis criminating people as nothing else does.

While so many of our egalitarian hopes were being dashed in recent years, the gentle revolutionaries were going right ahead, finding out how much you can do with shoestring power. It is they, individuals outside the universities, without Federal aid, who have pioneered the way to new forms of pollution-free energy for heating homes and running cars—doing it in isolation, as Steve Baer, among countless others, has done.

The crux of the matter *appears* to be power, but actually it is energy. Power is in the final analysis abstract. Certain people hold it because a lot of others want what the power symbolizes. I do not think that the weak must either wait around for the powerful to decide to share power with them—which is what doctrinaire thinkers believe—nor that their only recourse is to try to seize power by force, in which event, being weaker, they will probably lose. I am convinced that any time enough people simply turn away from those in power because they no longer covet that product, or mode of life, or source of energy, and build their own symbols of a rewarding life, they can peacefully drain and destroy the old power structures. This has been occurring.

The consuming interest of the so-called "earth people" who during the sixties wandered off into the deserts of California, Arizona, and New Mexico to experiment with the wind, the sun, triangles of plastic, and empty oil drums painted black, was what they called "low technology." This was not just a bland term for kinds of drop-outs housing and heating that would take us all back to the cave. But it was an approach that would put Everyperson back in charge of her and his own life again, fulfilling a deep emotional need of most human beings.

I have mentioned Baer because he is articulate as well as a doer. First, he chose to use the institutions rather than letting them mold him—going from Amherst to UCLA to Zurich, learning what he could but refusing to take a degree or become a conventional engineer, and just trying to get his own idea of what an education might be. Finally he wound up in Drop City, helping to build domes out of junked car tops. Then he and others solar-heated a dome with a big chimney stuffed full of rocks or something.

Later he originated a structural system he called a zome—combining zonohedron and dome—and insulated the walls with a mess of urethane foam, cardboard honeycomb, and dead air spaces. He and his family and friends heated this free-form structure with drums of water, employing the heat-storing properties of concrete flooring and adobe walls.

Baer told *Mother Earth News* (No. 22), "It's not very exotic or earth-shaking to fill 55-gallon drums with water, paint them black and place them in the walls of a home for use as solar collectors . . . But it works. It's a very low-technology idea that almost anyone can understand and use. . . . This is the kind of real innovation that actually makes organizations and keeps people happy. . . ."

He expressed the view, with which I concur, that the philosophical tactics and whole approach taken by huge corporations and governmental institutions, whether GM or NASA, miss the point of what people really want and need. And my own experience in higher education convinces me that its symbiotic relationship with the other arms of the power structure precludes much real social progress from that direction. Witness the academic neglect of Buckminster Fuller (one of the few genuine broad-gauge original thinkers of our time for many) decades following his dismissal from Harvard as a young faculty member. It remained for youth to discover his genius. Originality is threatening.

Baer mentions similar neglect of Harold Hay, California's pioneer in the use of solar heating, who gave Baer a radically new approach to the problems. But none of the engineers seemed to be interested in Hay because his ideas were so simple—the sort that could put an ordinary householder in control of heating his or her home. The guys with the degrees are "afraid people will laugh at them and say, 'What? You're a Ph.D. and all you're doing is opening and closing those doors?' "

There was dry laughter from our desert philosopher when asked if he had ever had Federal grants. Once he had been given free tickets to a National Science Foundation conference on solar heating. All they had wanted to discuss were projects that would cost ten million dollars before any heating or cooling got done—and which, worst of all, involved work that had already been accomplished.

Baer, in that interview a few years back, said, "It's going to be fascinating to see which direction society takes during the next ten or twenty years. It's really going to be fascinating."

And so it is. Big business has begun installing the first awkward solar heating systems on huge metropolitan office buildings. University architectural students are writing up yesterday's knowledge for tomorrow's theses. The system is being changed from the outside in.

And now I, it would seem, am proposing that the Federal government help those small farmers (or dreamers of small farms) whose very success has sprung from their initiative outside the system.

If you read the classifieds in the several periodicals devoted to organic farming, you will note that a huge population is seeking small-holdings. The young can seldom afford to buy. But they could if the government were to modernize the homestead laws, inventory both our public lands and our marginal private land, buy up the latter and resell it to homesteaders, make funding available, require that the land be improved so that taxing districts gain from its improvement; and *if* university agricultural extension services across the country were to pick up on the special needs of the small farmer, discover what is being taught in the ecology departments of their own universities and, combining that with basic farm knowledge, make a special effort to communicate with their new unimportant-looking clientele.

As more and more Americans have a chance to become miniature ecological systems, giving back what they take, we shall find fewer persons crammed into less-violent cities. The next time a nationwide strike of truckers occurs, fewer people in New York and Atlanta will fear starvation because their supplies from California have been cut off. The next time an oil and gasoline crisis occurs, fewer people in cities will be panic-stricken at the thought of not getting food, not getting to their jobs. The next time an electrical brown-out occurs in the cities, fewer people will suffer and there will be, in the first place, a reduced likelihood of such overloads. Fewer babies. And so on.

If we were to proceed with an up-to-date homestead plan, we should know in a very short time where any new towns would be needed—for the government would be *following* the flow of population rather than trying to predict or coerce it.

Former Secretary of Agriculture Orville L. Freeman, the author of a comprehensive study "Towards a National Policy on Balanced Communities," says, "We have space to spare if we use it. But we cannot use it properly if, in our planning, this space is constrained by the city limits, the county line, or the state border. . . ."

Local governmental jurisdictions need have no fear, however, for their powers and potential are becoming more important every day as this population reversal occurs. All the new settlers need—the poorest among them—is help of the sort our government extended to earlier settlers: homesteading assistance, credit, a means of expanding their knowledge and expertise, ways of developing equipment and marketing cooperatives. As they improve their land, Morton the Fork will be around in his pickup truck to reappraise it and the county will begin to earn several-hundredfold more in taxes than the old neglected strips were ever worth. Lest there be any doubt in the reader's mind, I am not proposing anything resembling the dole so generously extended by government to the huge agricultural conglomerates, the railroads, airlines, oil exploration firms, and to the mining and timber interests that have long enjoyed favored use of our public lands.

We must bear in mind that today's gentle revolution is being brought about by shoestring power. If government helps *too* much, or if any aspect of the so-called power system is allowed to internalize and co-opt it, we shall reexperience the imprisoning of ideas that followed earlier periods of innovation. A shoestring is a fine line.

It is important, therefore, to understand what is different about today's reverse flow, which will mean an entirely new character for rural America. In the past our farmers were born with roots in the soil, and they had limited access to cultural and intellectual centers.

Today's emigrés—many of them—have already learned about bureaucracy in the cities from which they come. At least as amateurs they are apt to know about grantsmanship and lobbying. Their education, creative talents, and craftsmanship will come with them. Their land skills will have to be learned. The end result will be a kind of rural culture we have not seen heretofore. Still important will be the county agent, the weekly newspaper, the county fair, the auction, the Grange Hall; but increasingly so will be the market cooperatives, craft fairs, small printing presses, health centers, community colleges, the nearest university campus, and cable TV. The extension of mass transit will enable more people to live on small farms while commuting to urban jobs (without unduly polluting the countryside enroute)—and this makes wise local zoning all the more urgent.

Good local zoning is consonant with—indispensable to—an effective Federal land-reform program.

The county where I live is illustrative of problems that every scrap of rural area in the U.S. must soon confront if land for small farms is to be preserved. In many parts of the country it is irrevocably too late. Each year 350,000 acres vanish from farm land on the fringes of U.S. suburbs; are removed from crops and put into buildings, parking strips, and freeways. The pressures are relentless, but they can be countered.

In our county, once the Benevolent Cement Company had closed its ancient plant at the entrance to the canyon and bulldozed down the old smokestacks, the cluster developers who had thus far successfully leap-frogged across much of rural California turned their attention our way. Cluster developments and big tract-mansion condominiums with security gates and inspired names like *Casa Safari Luau* (carved into what were lately living, thousand-year-old redwood trees) go hand-in-hand as it were—for the simple reason that they save money for the developers. A cluster development needs only a central access road and centralized utilities, and the multi-unit condominiums save costs for the same reason. Both are fine in their place. A cluster development may conserve needed land for parks if allowed near a city. Some people want the living style of a condominium. But people who need security guards and a life behind walls and conformity will find themselves extremely ill-at-ease among real rattlesnakes, real mountain lions and coyotes and bobcats and tarantulas. This is a basic flaw in the whole scheme that troubles the land developers not at all. Either one faction or the other is forced to give way—and that means, if you still have any wildlife in your area, the wildlife must go, except for the odd raccoon surviving on city garbage. By then the land developers also will be long gone—busily at work in the next rural county.

One thing we learned in our county was that just a few dedicated citizens, organized to promote good zoning, can fight the developers and win. It is a matter of perpetual watchdogging, however, because only battles, not wars, are won. The average General Plan has a life of only five years.

In zoning, the question is not one of either/or's. It need not be a matter of ranch vs. condominium, of 100 acres for a cow to graze on and 100 feet for a ghetto family to live in. Many lifestyles should be accommodated—that of the hip homesteader who merely wants a place to park his VW while patronizing the nearest rent-a-tree orchard, country store, or U-pick vegie patch; that of the retired frozen-meatball king who fancies a condominium with private golfcart path, and maybe a moat containing a plastic alligator. City people going rural may be genuine organic farmers, or professional people wanting merely a rustic retreat. Perhaps inevitably, planners have also begun to talk about something called the "homestead tract." Despite the name, which suggests conformity, a middle-class bankroll, and hence probably a stifling of adventurousness, it could be that the "homestead tract" will fill the needs of some land people.

Legitimate differences of opinion exist as to how much land is ideal for a small farm, as compared to how much one may really need to grow an adequate supply of vegetables for a household. I consulted city and regional planners for answers.

A city planner told me, "I wrote to four different agencies. To the U.S. Department of Agriculture, and the state agricultural agency, and to our county agricultural agent, and to the Davis campus of the University of California—asking each of them the same question. 'How much land is necessary for a family of four to grow enough food—exclusive of livestock—to support themselves?'

"I got four different answers. One agency said twenty acres; one said five; one said between one and three acres. And the fourth said, forty-five square feet of growing space."

I am quite prepared to believe it possible, with forty-five square feet of growing space, perhaps using hydroponic methods, and perhaps with the building up of layered beds, or various types of greenhouses and soils and fertilizers and methods of preservation, to feed a family of four.

A biologist at the University of California's Los Alamos Scientific Laboratory, working with algae and photosynthesis, reports that within a decade it may be possible to produce 3 million pounds of high-protein food a year from one acre of land—enabling the algae farmer to turn a profit of $1.5 million. This may represent a solution to famine, but certainly not to the desire of many Americans to enjoy the delights of country living.

# Chapter XXII

# Drought

In that first spring at Meanwhile a seemingly endless succession of bright, clear days brought anxiety—reflecting in miniature the mood of farmers throughout the Southwest. By the end of February, the swamp was as dry as a rattlesnake's eye. Occasional cloudbanks moved across the top of the canyon. I would think, now! Now it *must* rain. But the clouds always drifted on southward, and perhaps out to sea, leaving behind a dry world. I would never have believed that I could come to dislike the serene blue that first attracted me to the canyon.

These were months when California normally got several inches of rain to refill the underground pools and reservoirs for another year. It went against the grain to believe that blue skies could presage bad news, but literally, nothing was quite so damaging to grain. Crops that represented the total capital of farmers and ranchers were jeopardized. Range cattle found slim pickings on the burnt hills. Wells were already shallow, which meant little water for irrigation and that in turn, scant fodder for the following winter. It was all bad news and for me a jittery introduction to the farmer's lot.

One of my more unusual household pets during the early spring was a semidormant earwig residing between the glass and the screen of the bathroom window. The earwig did not eat, and it appeared to have little social life, but it got a terrific amount of sleep. Each afternoon if the sun warmed the glass, it moved a tiny distance. Sometimes it went South and hid behind the aluminum frame at the bottom. Or again, feeling an uncontrollable excitement, it might push North an inch or two. It became an important figure in my weather-watching.

On a farm there is always far more stuff that needs burning than a city environmentalist might guess—because you don't want to plow under your root-rot or blighted whatnot. I found that even after piling weeds and bark onto the compost heap, I had big piles of trash I desired to burn. I went to the State Forest Service headquarters to ask the ranger on duty for a burning permit.

A young ranger already embittered by red tape greeted me with "I'm in a big rush."

When I begged him to tell me all about it, he replied that he resented the inroads on his time of having to answer telephones and fill out forms when he had so many bigger, more manly tasks to perform, especially since, as he soon mentioned, he was a college graduate.

"And another thing! I'm getting tired of being cussed out by these farmers all the time. It's not *my* fault that every day thus far this year just happens to be a No Burn day."

"It must be very trying."

"The rules are all up in the air," he fumed. "The Air Resources Board is supposed to be working out new antipollution regulations. We have to wait for a radio call every morning from Sacramento to find out whether it's all right for farmers to burn. And then every farmer has to phone us after 8 o'clock to get the news. But the other problem is that the Forest Service shouldn't even have to be in this enforcement business. That should be done by the County Health Officer."

Trying to make a little joke, I said, "It sounds as if it's always a burn day for you personally."

He turned the color of his red tape and pushed some forms at me.

Looking them over, I said, "I have a new farm and want to burn only 'Small plots of grass or weeds in lots or dooryard premises,' and 'Small parcels or strips for hazard reduction.' That's all that applies to me on the form. But later I might want to burn brush on the hillside for a vineyard."

"You can't burn the brush," he said, not without relish, "and you can't do the other two things either."

On my form he checked two things he could let me do. One was "Burn trash in dooryard incinerator." The other was, "Burn orchard brush."

"I don't have an orchard yet," I apologized.

"Doesn't matter. You have to burn orchard brush. How much orchard brush would you estimate you will burn?"

I reflected that for a young man dead-set against red tape, he was falling into the ways of bureaucracy with flair.

I said, "Dang it, how would I know?"

He wrote down, "One ton of orchard brush."

He also wrote on one of my forms, "Permittee has adequate manpower and equipment to keep fire contained."

I assumed that meant the farm dog Alice and the garden hoses she had been chewing up in the early phases of puppyhood. (It was a "methadone" period for her: I was trying to wean her away from chewing up small trees and rose bushes with this daily ration of garden hose.)

"Now, call me tomorrow morning after 8:30," the ranger said, "and I'll tell you if its's a Yes Burn day or a No Burn day."

He forced a smile around clenched teeth.

"Could you just give me an idea," I asked, "about what the chances are? I mean, would I be apt to get one Yes Burn day a month, or two— or any at all?"

(The basic catch to the system was that, assuming one were lucky enough to strike a Yes Burn day, it was almost bound to mean rain, and you would have to race against your brush becoming too soggy to ignite.)

He shook his head. "No way to tell. Weather like this, you could call me every morning for a month without striking it lucky. And about the fifth day, you're going to start cussing me out and blaming *me* for it."

"Don't you *like* having Acts of God laid at your doorstep? Just a teensyweensy bit?"

"No, Ma'am, I do not."

A radio began chattering in another office and he dashed for it. I decided I would give God about five days before taking matters (and a few matches) into my own hands.

Shortly after this dramatic encounter, I happened to be checking my household earwig and found that it had zoomed to the Southwest and vanished beyond the window frame. A new position, clearly a departure from old movements. Curiously I studied the eastern sky where the sun was just coming up through soft, rimpled little clouds the color of red salmon.

I noticed that the weather had a sultry feel to it. It looked and felt like—rain!

I dashed to the living room and tapped the barometer, which never worked.

I flew to the telephone and dialed the ranger's office, trying to control the trembling of my lips.

He answered the phone in a voice exultant and triumphant, crying, "Yes! Yes it is indeed, it *is* a Yes Burn day. Go right ahead! Touch a match to everything in sight, provided it's orchard brush."

"I can hear the rain beginning to patter on the roof right now," I said. "And do you know what? I *knew* it was going to rain because—. Never mind. Thank you."

I had meant to say: "Because my earwig moved Southwest." But why waste breath? He would still say it was Sacramento that had made it rain. And besides, I couldn't be sure yet. It was only the first time. The ranger would be bound to point out too that I didn't even have a control group.

It pleases me to report, however, that subsequent observations verified that a direct southwesterly dive meant *light precipitation*. I quickly named this loyal and reliable insect Yes Burn. And I never again experienced anxiety about when to burn my orchard brush.

The only thing I felt anxious about, as the drought resumed after a single day of showers, was whether I should *ever* get an orchard planted in that hard, unwelcoming earth.

# Chapter XXIII

# Mother Nature's
# Seamy Dirndl

The warm, dry spring alternated with heavy, crop-killing freezes. I marvelled at how quickly after one of these the diabrotica came out, tumbling in lavender and green cascades along the creek bank, reminding me of the long, idyllic childhood weekends at Combray . . . of dipping madeleines in Valerian tea . . . the good Morel (a ham, a mushroom?) . . . and of course invariably the Verdurins.

Spring has always been for me its own hallucinatory. Diabrotica is—or are—a common cucumber beetle. But like so much else that is seamy in Mother Nature's dirndl, the lyrical sound of the word, I thought, deserved a higher destiny. Proust would have welcomed the diabrotica to his cast. In my omnivorous winter reading I had veered heedlessly from the outpourings of the Rodale Press to Proust and return, including Ruth Stout in the former category, until I scarcely knew one from the other and was content to reap my little harvest of confusion and put it by on the preserving shelves of memory.

The crazy rush of seasons left no time even for alphabetizing my seed packets as Ruth Stout insisted one should do, or for drawing maps of the future garden layout as Rodale did. The race was on. One could not hope to cope with spring, but there was joy just in showing up for the encounter.

One morning I was awakened by the sound of light rainfall and leaped out of bed—but not quickly enough to catch more than the very end of the little storm. In a moment blue ink was again spilling after white egrets, chasing them across the canyon sky, and the drought had resumed. Little breezes began to lift straw and leaves from the raw garden plots and spin them into the air.

Then more clouds hurtled across the ceiling, now black and glowering. Sunlight flared from behind them, casting a new perspective on the peculiarly graceless, bone-white beauty of the sycamores. This was the beginning of the first true spring day of the first growing season since the earth was formed.

I looked out of the French doors, down the length of the canyon to the West, and was surprised to see the ancestors of dogs prowling hungrily through the swamp for mice, seeing for the first time this world in which human beings played no roles, influenced no trends, followed no opinion-leaders, left no footprints in the mud near the watering holes.

Yet this was a world that had existed, adequate for the most part to the needs of its creatures. When this prehuman land had been inadequate to the needs of its creatures, there had been no critics to say or even to think that things ought to have been different. Many had died young and that was that. The earth, the canyon, the swamp, the hills, the waiting earthquake vault, and the rather large hairy animals prowling for mice—this scene, these players had had no need of me or of the things that my kind would deem essential. For them, no such creatures existed. The God of wild dogs had created wild dogs in His own image. (It was only Alice, of course, chasing about with a dog from a neighboring farm.)

I went out before breakfast and, in thoughtful mood, planted two-and-a-half rows of peas, chosen largely because they were Early Dwarfs and I counted on the weeds being regularly pulled by Snow White and her troupe. In a spatter of rain, I finished, and ritualistically slipped the illustrated seed box down over a stake. This ceremonial act was one of the great moments in gardening, the instant when, after straightening an aching back, brushing hair from eyes with grimy fingers, one stooped again to add the graphic promise of perfection, knowing that it might well be as near as one would ever come to it. God had never promised me a peapatch.

As I was performing the ceremonial act with the seed packet I heard wild screams and straightened up to see three redtailed hawks circling against the hillside, skimming down over the tops of bushes. Their cries might have been for exultation at good air currents or redtail courtship or simply hunger. It occurred to me sadly that the hawks had no sense of their own great beauty and that the word symbols in my head (wild, free, beautiful) were for them a false concept, a human one, and that the reality for the hawks more likely would read *fierce hunger in the belly*. I thought of the university scientist, enlightening his television audience about raptor *management*. In terms of the hawk's sensibilities, I could never know the reality. Why could I not cope with the scream in its pure state, absorbing the emotion unlabelled as my dog probably did?

I wondered, for example, about Alice and the koi fish. The koi is a Japanese carp about a foot in length, speckled gold, white, and black, that lives in the pond. Such fish may live to be five hundred years old and the Japanese call them the Fish of Wisdom. They are very tame and will let you feed them by hand. I have not taken time to make friends with our koi, but I try always to throw its feed near the bank. Alice and the koi spend a great deal of time looking at each other through this stretch of water. I mean really standing/floating, looking into each other's eyes; and I wish I knew what it was all about. I doubt that they are anthropomorphizing.

I watched the three hawks until they floated high up among the clouds and vanished.

In a sense I was resisting the planting frenzy, feeling that I had not even thoroughly absorbed the look of winter in this broad, flat farming country. Spring fever filled me with lassitude. I balked at being harried and spent hours restlessly wandering in and out of doors, starting tasks, dropping them, making excuses to drive off on errands when I merely wanted to look at the young livestock being born in pastures along the roads. I had seen the first lamb in the canyon, the first colt, and a great many wet-nosed,whitestockinged Hereford calves.

The little range calves born into freezing weather were not only tough but absolutely secure in the conviction that their mothers could protect them from anything—which, alas, was not always true. Occasionally a pack of wild dogs or a mountain lion would take its toll. With high beef prices, however, rustlers were even more likely.

Visitors to Meanwhile, hearing a range cow utter its strange and lonely bellow, tend to start and say, "What's *that?*" It is hard, having been brought up to believe that cows say *Moo*, to credit the expressiveness packed into the voices of these half-wild creatures. The cows have stocky bodies, short, white, rather smug faces, blue eyes with white eyelashes, and small, classic udders. They trot along the hillside paths, usually with one beautiful kneesocked calf behind, but occasionally with twins. If a large dog challenges them, the mother lowers her head and stands menacingly still while the calf moves close to her side, showing no fear.

To enjoy a final glimpse of winter-drowsy earth, I went for a drive through the surrounding countryside, overtaken by every impatient rancher on the road. Everything looked brazenly sensuous. An orchard of Comice pear trees as formalized and tapering as well-girdled matrons, swollen with spring buds. The Almáden hillside vineyard curved like the Earth Mother's breast, with recently pruned vines forming neat arches of red and green shoots.

I was already too late for winter. Certain fields showed two inches of bright green—spinach, lettuce, garlic, and the first of the new flowers grown by the Ferry-Morse Seed Company. When the hot months came, they would form shimmering rectangles of color.

Back at the swamp the spring nights had turned brilliant. A single frog now croaked bravely in the pond. I was glad for its voice during the hours before dawn since the owls had temporarily deserted. This lone precursor of frog choruses had a sweet, clear voice.

There ensued days when the sky was heavy with unshed rain or with more of the fine, false clouds. The local newspaper reported there had been no drought of this magnitude since 1856 (why was 1856 the year of all great records?). I could no longer delay planting an orchard.

Taking it in small measures of pickaxe vs. hardpan, I put in first two grafted English walnut trees, a Red Delicious apple, two Blenheim apricots, a Van cherry. My order was in but unfilled for a green Pippin apple tree, producer of the shiny, tart, elliptical, bright-green fruit that contrasts beautifully in a bowl with Red or Golden Delicious. The poetic names of trees and vines, like that of the beetle diabrotica, I found subconsciously motivating my selections.

The order for a Green Gage plum tree was inspired in part by its sensuous, pale-green fruit, but also by Rumer Godden's charming novel (*Green Gage Summer*) about a family of English children who found themselves stuck all of one summer in a fusty French residential hotel. Liberated by their mother's confining illness, inspired by the hotel's old walled garden and the surrealistic glow of the plum tree, they managed some marvelous escapades. The youngest child was a five-year-old prospective transvestite who particularly took my fancy. Little Marchmont, if I remember his name correctly, was preparing for a career in haute couture. Tagging along after his older sister, he invariably carried a copy of *Harper's Bazaar* under his arm because his hands were tucked into his ermine muff. The eldest daughter got mixed up with a gangster. All in all it was a very rewarding summer. I shall think of them all with gratitude when the first Green Gage plums have ripened.

I planted little almond trees which thrived, and a fig that the gophers ate, and sugar prunes that the gophers ate, and more of these and other varieties.

The first grapes I planted were table varieties: Thompson's Seedless, which are small and green; Ladyfingers, which are long and cool and seedless; and Tokay, the heavy-clustered russet grapes that supply the opulent look and taste traditionally associated with a vineyard in autumn. The following year I managed to afford a few varietal vines, beginning with the Pinot Noir from which Burgundy and champagne are made. Again the visual appeal was important. The skins of these grapes are of a vivid, dusty, blue, their clusters Bacchic. In Van Gogh's paintings one may see the colors of the inside of the skins, a contrast of red-blue which on his canvases achieved unequalled brilliance. And then I found two vines of Cabernet Sauvignon—not much to start a vineyard with, but at least I dreamed big dreams.

As I planted the last of these vines, the earth began to shake in a rather unseemly fashion. It would not do to leave such important work unfinished so I hurried, and puffed, tamped down the earth, and added water from the hose. A few more shakes occurred in the late afternoon.

With drought already, I needed an earthquake?

Chapter XXIV

# Acts of God and the IRS

If it is true that nothing is more certain than death and taxes, April brought a hint of destiny to the swamp on both scores.

I was awakened late that night by the house trying to knock itself down, in part using my head as a battering ram against the headboard. I heard the curious roar that people who live in earthquake country grow to associate with sudden terror—a windy rumble that resembles no other phenomenon. Afterward one always felt unsure whether the sound came from beams, boards, and foundations grinding or from the deeper protests of the earth. To this sombre theme, one heard the dishes dancing on shelves. In my bedroom, the tiny prisms in the "Earthquake Machine" swayed wildly.

Clawing my way out of bed, I lurched toward the door, for I tend to be claustrophobic. Then I paused, remembering that most injuries in earthquakes resulted when people started rushing around. They were crushed on the head by beams or slivered by flying glass. After a moment the house stopped jumping. I returned to bed and pulled pillows over my head.

Seconds later another jolt struck, much harder than the first. This one made me mad, plain mad, the way I would feel if someone deliberately stepped on my foot. What had I, a friend of Earth, done to deserve this brutal treatment? It was unfair. Again I prepared to rush outdoors where, if need be, I could die feeling a little better.

But gradually on that dark April night in the swamp, the earth settled down. I went back to sleep. At around 3 o'clock in the morning another violent quake awakened me. A few aftershocks occurred, which ended the business for then.

Next day's news reported it had been a five-plus on the Richter Scale, which is "major." But there had been little damage in the area.

The earthquake jolted me into reexamining my motives in moving to the area. In fact I had put out of my mind a danger of which I had been well aware. Everyday life made fatalists of us all, all risk being relative. Unquestionably, as I now had to admit, some risks were more relative than others.

I had begun to understand how people felt, however, who lived on the slopes of active volcanoes and who refused to abandon their homes even when driven forth time and again by hot ash and scalding lava. I had begun to understand how Nicaraguans could begin rebuilding Managua on the ruins of its original site. This was their place in the world, however it might treat them. President Lyndon Johnson, writing of his Texas hill country, touched on this, on how people who lived on a land became part of it and were shaped by its qualities: "So this is our home. We take what we can from it—as our ancestors did. But like our ancestors, we always seem to have to give more than we take and after a while, the act of giving becomes an act of respect and finally, an act of love."

A few miles from my home one may read a somewhat bizarre National Park Service plaque declaring, "San Andreas Fault has been designated a registered natural landmark under the provisions of the Historic Sites Act . . ."

One might dismiss such zany chauvinism with a pun as simply a case of bureaucrats being generous to a fault; but one should also accept it in the spirit in which it was offered—a macabre obeisance to the power to destroy. We are at heart great respecters of such power.

In harmony with the season I had been reading with unusual zeal a pamphlet prepared by the Internal Revenue Service to help us farmers make out our income tax returns. When I first arrived in the canyon, my neighbor the well-witcher had enlightened me as to the crux of this matter.

"You can't deduct taxes for your heavy farm expenditures like land drainage, pond-building, damming, weed control, or other reclamation work," she told me, "nor on investments in orchard planting, or even burning your orchard brush, *until* you have *earned a profit from farming.* To earn a buck is one thing; for it to come out as a *profit* is another."

She had a disturbing habit of plunging straight to the bleeding center of an issue. I looked around me, thumbed over my Banty eggs, and saw no profit. You can live well on a farm without a whole lot of cash; but *profit* is just about as ephemeral as the Mayfly.

As I came to the section in the IRS pamphlet entitled, "Involuntary Conversions," however, a tiny light began to flicker in my head and I jerked up in my chair, breathing hard. The deeper I read the more clearly I could see that I ought to just sit right there in the chair, reading the fine print forever. It was here the profit would be made—not rushing around in the mud with my spade and my egg candler.

"If your property is involuntarily converted into money," the report stated solemnly, "you may, under certain conditions as explained below, postpone paying tax on any gain realized on the conversion. . . . Livestock that die from disease, or are destroyed on account of disease, *or are sold or exchanged because of disease* (italics mine), are involuntarily converted, even though the disease is not of epidemic proportions. . . ."

One inference seemed to be that if you really wanted a tax breather you shouldn't fool around selling just one or two sick animals; it would be the smart thing to have an epidemic. There was, however, much more—including the slightly dampening intelligence that, "Sales of poultry . . . because of drought are *not* involuntary conversions." There was bureaucratic arbitrariness for you.

Some of my new neighbors had bought calves at the livestock auctions, only to have them die in a week or two of infectious disease. Little had I or they suspected that the government rewarded the sellers with tax advantages. Apparently one could sell a sick anything as an involuntary conversion. Uncle Sam was certainly leaning over backwards to give us farmers a break.

Farmers, like lawyers, politicians, and people in other occupations, sometimes create unlikely enterprises that are intended to lose money. These are widely known as tax shelters. With a little bit of luck you may get away with charging your farm gasoline or some other expense to them. Anyone driving along a country road who happens to see, smack out in the middle of a hundred acres of rutabagas, a sign advertising the Laser Research & Palomino Stud Service Facility, would perhaps be justified in suspecting hankypanky combined with long, deductible business lunches down at the Bitterwater Cafe. I myself am constantly on the alert for some such enterprise, just *in case* I ever make a profit at Meanwhile.

A farmer I know told me he was thinking of combining the raising of rabbits with growing Christmas trees as offering attractive long-range tax shelter potential.

"But—won't the rabbits eat the young trees?" I asked naively.

Chuckling, he asked whether I had not read the section on the involuntary conversion of a Christmas tree into a rabbit.

Anyhow, that's mostly what we do out here in the country on cold nights in early April when, might be, we're waiting for another 'quake. We're sitting around the Franklin stove in our bedcaps and nightshirts, with our feet on the hearth, like Beatrix Potter's mice, mulling our applejack and squinting at the fine print in every pamphlet we can get from the IRS.

# Chapter XXV

# Frog Slither

One spring morning I awoke before daylight and went outdoors to sit on the front steps. Alice rubbed against me, gave me lickies, and put her nose into the air to sniff at the blurred new moon balancing on the edge of the opposite hill. "Smells good, hmmm, Alice?" I said, just to be sociable. She went on sniffing appreciatively all up and down the sky.

I closed my eyes, listening to the birds waking up with their sleepy different kinds of sounds. Then I looked around the pasture swamp that was now almost a farm and began to feel a surge of wild, ungovernable, sinful pride. It had been only a matter of months. I remembered how high the weeds had been before Entwhistle's bulldozer. And now look at them! Almost as many as before and just as robust. Nothing could kill this swamp farm.

Even if the little orchard trees began to grow, it was still going to be the same old swampy hollow with the same old tarantulas lurking around, waiting for the rain so they could get across the road. A tarantula *hates* a drought. Makes it so hard for him to get anywhere.

Look at those flower beds—hardly any of my seeds had managed to struggle up, yet the wild yellow nemula, the goldenrod, the lavender-and-green mint, and the tiny salmon-colored blossoms from which gophers made their salads (hummingbirds and deer favor red blossoms but gophers like that salmon shade)—those hardy volunteers had made it.

Out on the county road in the morning gloom two pickup trucks, travelling fast in opposite directions with their headlights on, passed in front of the mailbox, the drivers' hands flipping up like automatic signals.

I had come to the farm feeling down on prospects for the human race and not wanting more to do with it than I could help. But now, oddly, in a few months of hard work I had begun to feel optimistic. For the first time in a few years my own life felt as if it had purpose, a fact which I blissfully assumed must augur well for the entirety of mankind. Good luck world, I thought. Maybe you are going to make it after all.

Because of the drought there was fear among the growers that a killing frost might strike after the fruit and nut trees had been forced unseasonally into bloom. Nature began to seem like one hell of a conspiracy. Other bad combinations were too-hot weather just at ripening time that would cause the fruit to drop, or untimely rains that interfered with pollination.

I continued to learn new things about country services.

In this region we had both an itinerant blacksmith and a temporary U.S. Weatherman. The latter, as a service to fruit-growers, took up residence for a month or so each spring, expressly to warn the growers by radio when to fire up their smudge-pots. He advertised in the local newspaper: "Wanted, house, six weeks, family of three, 2 bed., by Weatherman." Anywhere else such an ad would have had the FBI on his trail.

Our itinerant smith tours the country roads, calling by appointment on people with horses. He is a colorful person, his forge a smoking stove in the back of a pickup truck, something like the reddleman in a Hardy story; but with the growing popularity of riding, he is no longer an anachronism. In fact his calling has been described as one of the little-known but keen job opportunities of the future, comparable to chicken-sexing insofar as arcane knowledge and skills required, although I doubt that it pays as well.

We also have an itinerant slaughterer who will shoot your steers, butcher them, and package the meat for freezing.

Our weatherman took up residence none too soon. I awoke one morning to find the whole sky a greasy black, the result of oil smoke from a myriad of little smudge pots (a method no longer permitted). But even so, as it turned out, the apricot growers lost half of their crops.

My dry vegetable garden looked dismal. I made a liquid fertilizer in a big plastic garbage can by turning the water hose onto horse and cattle manure. This I emptied onto the pathetic little struggling plants.

I began to seize on the smallest successes for emotional sustenance. The sight of little grapevines pushing forth crinkly, shiny, delicate leaves, carrying within them as they did the promise of vigor and future ribaldry, raised my spirits.

Alice one morning failed to respond to my call. I found her up to her armpits in the pond, engrossed in the activities of the frogs' nursery. As baby frogs were born she spent each day patrolling the bank or simply lying in the grass with her chin on her paws, an utterly frog-bewitched dog.

Swarms of wild honeybees had emerged from hibernation and were buzzing thirstily all over the water. In the grass, what looked like fire ants were feverishly locked in mating clinches. A ladybug impetuously leaped into my blouse. And as Alice and I were gathering watercress in the ditch below the dam, we heard something that we never told to anyone else. Two frogs were *whispering*. They knew we were there so they whispered across the pond to each other several times. On the grassy bank I finally saw one of them, a tiny green creature with darker green patterns on its back and golden eyes.

The bird population was suddenly swollen by exotic summer callers. A pair of Western tanagers, brilliant as parrots with their red faces, passed through. A flash of yellow, orange, black and white among willow leaves disclosed the presence of Bullock's orioles. A pair of blackheaded grosbeaks, their russet breasts trimmed in lemon-yellow piping, built a nest in an elderberry tree near the pond. Swallows, Western bluebirds, gold finches, house finches, redshafted flickers, hummingbirds, marsh wrens, and various kinds of flycatchers—all common to the area—became frantically absorbed in nesting.

The air had begun to smell of the ripening wild hay, the mint, and marsh grasses—a rank, swampy smell, streaked with sudden whiffs of sweetness or pungency, a nasal feast. In deep leaves near the creek I saw a slender, motionless, brown-and-yellow snake. Since I was unable to see the head, I could not tell whether it was a young rattler or a young gopher snake. Occasionally one also saw dramatically banded black-and-white king snakes.

The quails and mourning doves raised their cries from hill to hill, and the mother skunk led her kits forth in a straight line of tails from the palatial bramble patch. Here at least for a time a place remained that belonged first of all to the wild creatures. And because it belonged to them, we could praise the music of human joy:

"Praise we the Gods of Sound—
From all the hearths and homes of men, from hives
Of honey-making lives:
Praise with our music those
Who bring the morning light
To the hearts of men, those households of high heaven!
Praise. . . ."*

---

*"Praise We Great Men," by Edith Sitwell, for Benjamin Britten.

# Chapter XXVI

# The Manic Organic

The liveliest war of words raging these days is between organic gardeners and the champions of agribusiness—the purists vs. the self-styled scientists and technologists. Ruth Stout, the peppery New England octogenarian, seldom even uses such words as *scientist* or *expert* without putting derisory quotation marks around them. Her bias is understandable. Higher education, with its generous research grants from the manufacturers of chemical herbicides and pesticides, was extremely slow to give students the benefit of organic gardening methods. In land-grant universities, especially the older, urban ones, organic gardening has still not commandeered the main agricultural hall but has crept in through the back door as a rule, via the garden plots of married students' housing. It is taught usually by graduate students, at the strong insistence of other students, and despite great popularity is still considered peripheral. At new, ecology-oriented campuses such as UC Santa Cruz, however, organic gardening was encouraged from the beginning.

Ruth Stout advocates selling all your spades, hoes, your plow, and your cultivator at a second-hand store, and giving up weeding, composting, plowing, spraying, and hoeing in favor of just spreading a year-round mulch of hay on your garden.

My initial response to her message was, "Who wouldn't go for a no-work garden?" Yet as I thought further I realized that the idea ran counter to much that lies deepest in the puritan character: the drudge ethic, for starters. But it was also deeply offensive to those who worshipped in the temples of horsepower, i.e. 96.3 percent of all men. And still another group must be repelled by the Stout method—those who on principle prefer complexity to simplicity because mystique may be translated into arcane knowledge which in turn they are in the habit of translating into power.

Mrs. Stout accepted nothing from nobody in the gardening line, unless she herself had tried it in a systematic, replicative way and found that it worked (which makes her more of a scientist than some she might care to mention). I know of no doctor of nutritional sciences, for example, who like Ruth Stout lived for a full year entirely on uncooked food simply to learn whether cooking might be just another of those needless, expensive, time-consuming frills. And who did so not as an exercise to fulfill degree requirements but out of the genuine desire to know and to advance human knowledge. Nor was it food-faddism. No special food was bought for her diet during the experiment but she ate raw whatever her family ate cooked. Apparently her experience proved that for her at her then age and hence, no doubt, for many others, cooked food was *not* essential, for she says that her health remained good and that she neither lost nor gained weight during the experiment.

In the *No-Work Garden Book* which she wrote with Richard Clemence, she cites an issue of a Rodale Press magazine with at least eight articles about gardeners who, in her opinion, worked harder than necessary to grow good crops.

Here was a man who did wonders with his compost, and there was a picture of a *ton* of the stuff ready to be distributed. What Mrs. Stout sensibly wondered was whether he would be able to manage all that work when he got a bit older. All you really *needed* to do was put the leaves and hay and refuse directly onto your garden and let them decay on the scene.

Her "scientific" opponents claim you can't build up enough heat for decomposition this way and that lingering garbage microbes may infect your new produce; but Ruth Stout, a pro to the core, simply ignores such niggling.

Then she reports amusedly on a female organic gardener who, to stamp out cabbage worms, first employed "trap-crops without much luck, then tried spraying the plants with salt . . . Then she tried dust, then rye flour . . . an onion-garlic spray . . . She finally settled for sour milk . . . but she admits that it doesn't get rid of every worm."

By this time the reader is down on her knees, beseeching Mrs. Stout to disclose her secret. How does *she* dispatch the cabbage worm? To begin with, thanks to her simple, labor-saving, year-round mulch of hay and garbage, and perhaps because of her formidable convictions, bugs seem to trouble Mrs. Stout very little. (And, incidentally, Robert

Rodale too swears that insects will give little trouble to plants grown in healthy soil.)

"Twice a season, or possibly three times," Ruth Stout reports, "I go down my cabbage-family row and sprinkle a little salt from a shaker on each plant." Almost gratuitously she adds, "This has been my procedure for several years, during which time I haven't seen one cabbage worm."

"Scientists" and other low skeptics of an academic hue often go to visit the Stout garden and, with her permission, poke around, seeking flaws in her system. Sometimes they write unflatteringly of her, referring to witchcraft and faddism. All this in vain of course since it merely adds powder to her literary flintlock.

It may have little to do with Mrs. Stout's contribution to world nutrition, but somehow it pleased me when she mentioned also that her husband, Fred, kept turtles behind low fences.

I was already a Stout convert when I planted the first garden at Meanwhile—and I decided to give the mulch method a three-year trial. But compacted soil like mine needed air and spading too—even she acknowledged that. Her own garden had been under cultivation for thirteen years before she began mulching, so all her soil needed was the simple care she gave it. I throw refuse and straw on my garden, but once or twice a year Angelo comes with the *machina* and works it up. Another decade or so, and I shall perhaps write a book entitled *A Sloucher's Guide to Mulchery.*

The idea of year-round mulch came to Ruth Stout through noticing one day how a messy and long-neglected asparagus bed was thriving under its blanket of decaying weeds and its own dead tops. And I found this fascinating in the light of my own experience with asparagus. When I had lived on the farm in England, an Englishman solemnly told me, "Here we say that there are only two things a man can leave his son—a college education and an asparagus bed."

The reason for the gravity of asparagus in England was the belief that it took eighteen years and God-alone knew how many man/mile hours of trenching, fertilizing, sand-hauling, praying, cultivating, and stem-mounding before a harvest could be hoped for. With lightning calculations I figured that a young Englishman would have to be a slow learner to make his education come out even with the hereditary asparagus crop. When I returned to the United States, the first gardening catalog I picked up told how to grow asparagus in *two* years.

If Ruth Stout is to be believed—and the idea of not doing so terrifies me—the way to grow asparagus successfully is to throw a few of the roots down on top of some orchard grass, pitch a forkful of hay on top of them, and go away to salt your cabbage worms for a couple of years. Your son may wind up studying automotive mechanics at J.C. but if these asparagus roots do not burgeon into the most delicious crop you have ever eaten, please write to Mrs. Stout, c/o the Rodale Press, with full particulars.

Richard V. Clemence, her collaborator, says that nothing in his experience of the Stout System, nor in that of any other gardener he knew of, indicated regularly larger yields or better produce than other organic methods. And he points out, moreover, that Ruth Stout herself has never made such a claim, "despite the fact that many people seem to think she has. . . ."

And that is characteristic of the keen spirit of the organic movement. A typically misled correspondent writes: "Ruth Stout is always roaring about her year-round hay mulch . . ."

If she threatens our emotional needs for martyrdom or our intellectual needs for complexity, or would simply deprive us of the pleasure of spending long hours muddling around outdoors and having a virtuous excuse for doing so, she is at least in her own case able to justify all this by the fact that her gardening methods have left her time to write several books about no-work organic gardening which, presumably, have earned more money for her than she might have made by operating a roadside corn stand. And that's good enough for me. It beats burying the ashes of a rejected manuscript among your cabbage family and having it come up worms.

# Chapter XXVII

# The Violent Month

A few days of heavy mist brought slight relief from the drought toward the end of April. In the caked swamp, wild mustard, thistles, nettle, hemlock, and all the other rank opportunists promptly surged. Huge nemula, bright yellow with a sprinkle of cinnamon on their tongues, filled the ditch, contrasting with the tiny white blossoms and dark green leaves of watercress gone to seed prematurely.

A new gardener's humiliation occurred: at Easter I had gone on a trip after carefully transplanting little bell pepper plants from a cold frame to the garden. Returning a week later, I noticed proudly that not only were they thriving but that there seemed to be twice as many as before. Bumper crop of bell peppers! But what were those tiny needles below the first row of leaves? And *why* didn't the crass seed merchants give amateur farmers some clue of what seedlings looked like instead of using only those lush likenesses of mature vegetables on the packet?

I dialed the county agent in my usual panic.

"Don't know what you mean by little needles," he said. "A pepper has kind of oblong, deep green leaves. Tell you what, though. A bell pepper takes a long time to germinate. Several weeks."

I thanked him, slammed down the receiver, and raced back out to the cold frame. Sure enough, I could see tiny plants just poking through the earth. Obviously I had transplanted into the garden a great number of very healthy thistles.

Worried that nothing was growing, I treated the vegetables to intemperate doses of chemical fertilizer. This caused leafy plants to turn a uniformly deep emerald green as a result of the nitrogen. I proudly gave a friend a sack of spinach only to have her tell me later that it was lettuce.

The pond was now filled with baby bullfrogs that floated with their legs asprawl and their mouths dreamily cushioned on bullfrog chin-bubbles, while other tiny frogs, presumably female since they lacked hot-air facilities, floated nearby, staring worshipfully into the faces of the young croakers.

Just before sunset each day a period of intense excitement gripped the swamp creatures. Flights of tiny goldfinches and flycatchers whipped noisily through the air in pursuit of bugs. Starlings added their peculiarly decadent sucking sounds. And suddenly every living thing was singing, the notes bouncing back and forth from the canyon walls, while the sad cries of the mourning doves tolled a minor theme. The young bullfrogs would raise their erratic music. Alice would wander off to the pond, looking back toward the house at intervals. The most striking aspect of this evening performance was her responsiveness to the mood of the wild ones.

Even Ferd Farkle responded in a cool sort of way to the sunset frenzy and would venture into the field. He returned one evening distastefully carrying a rodent that I identified in a book as a large shrew. According to the book they had poor eyesight but an excellent sense of smell, ate voraciously, and were highly nervous and irritable. Thus I came upon a description of the shrew mole or *Neurotrichus gibbsi* whose habits and description charmed me. An obvious eccentric, this unprepossessing creature had a bare, turned-up snout which, when walking along, it tapped the ground with—perhaps to detect insect vibrations.

Each morning was heralded with nesting invitations and territorial proclamations. My own day began when I opened the back door and Alice, holding her empty chipped green saucepan by its handle, would rush in so bumptiously that the force carried her straight through to the far side of the living room, she and the pan crashing against the wall in a heap. She would pick it up and rush back, ostentatiously flipping it upsidedown at my feet so that any possible remaining dregs of kibble would spill onto the kitchen floor. Ferd Farkle, at Alice's entry, always leaped for his cabinet, the glare in his blue eyes explicit: a dread, cat-eating monster had broken into the house, bringing several infectious diseases.

Eventually I had to start chaining up Alice's saucepan to the back door like the librarian's scissors. The trouble was that Banties and ducks started hanging around to steal her grumblies. If she wasn't *ready* to clean up her food, she would have to carry the pan everywhere with her, which got to be a nuisance. She couldn't be watching it *all* the time. Her solution finally was to carry her old green saucepan with its remaining crumbs out into the swamp and bury it in her very own root cellar. That way, if she woke up hungry in the night she could always find it. But I could not, and Alice seemed to forget. I lent her a series of mixing bowls after the green pan vanished. She carried them all away and buried them. Her habit was becoming expensive and I was growing short of utensils. When one of my good aluminum saucepans vanished, we stopped the whole business with a chain. Perhaps some future archaeologist will hit on Alice's kitchen midden and theorize a buried city populated entirely by a very backward breed of cake-mixers who of course ate only with their hands.

Suddenly April, the violent and beautiful, the seminal and germinal month was gone. I walked across the creek one morning, moving quietly and feeling quiet inside myself. I wished I might always feel as quiet. The season was hurrying into summer, the creek shrunk to a trickle of brown iron sediment that stained rocks, mud, and last autumn's rotting leaves. I walked on up the steep path beneath oak trees, skirting the poison oak jungle, pausing near a stump to watch a lizard shed its tail. A brown towhee scratched leaves. Coyote droppings.

The horse chestnut at the top of the draw had just begun to flower. Some of the spikes had a white and slightly ragged, orange-stamened, small blossom just at the end. Glancing up, I saw a brilliant butterfly, wings transparent against the sun, ripping savagely into the flower. I picked one of the blossoms, finding that it smelled of mock orange; and some sprigs of spicy California sagebrush, and sticky yellow monkey-flowers, and the leaves of wild roses not yet in bloom, and the flat, oval, tiny leaves of a nameless bush that smelled of green metal. These fragrant gatherings would be put into a large manila envelope and mailed to Dr. Yee, who was again in a hospital as a patient. This time she had elected surgery that offered a fifty-fifty chance of survival, because, if the right fifty came up, she could resume travelling the world. Which, I am delighted to report, it did, and she did.

The hillside was warm and quiet, the way I felt inside myself. I never wanted to hurry again, nor to be driven by things and urgencies, either of my own making or others'. In fact I had been moving slowly and feeling peculiar for about two weeks.

On up the hill a doe appeared leading twin fawns. Alice barked. She turned quickly and vanished with her family into the ravine.

The previous autumn, long after hunting season closed, poachers and deerjackers had continued to blaze away. Not a morning or evening passed for months when shotguns could not be heard. Some said that a certain distinctive van in the area belonged to "gun freaks" who were selling venison. Whatever the case, in the current year the deer were few and far between. Does as well as bucks apparently had been wiped out. Local ranchers tried to preserve their deer breeding stock, but the hills in the canyon were the property of absentee owners.

As I returned from my rather somnambulistic walk, a mist began to fall, cooling the air and perhaps softening the earth. Thus, on this day of listlessness and flower fragrance, I began weeding the garden. That was how I happened to see a movement in a gopher hole. It proved not to be the primary resident, however, but a benefactor—one of the ploppy, engaging, old fat toads that hung around, brown and maroon with huge splayed toes, looking like overly tossed bean bags. The fat toad seemed pleasantly surprised by the fact that I had inadvertently turned a water sprinkler full on it. Toads must drink through their skins. I felt guilty that I had not set out a pan of water for such an important assistant—a gobbler of ten thousand bugs a month according to one amazing specialist in toad droppings. I had seen a toad at dusk join up with a partner and set out for a bit of hunting, uttering curious birdlike calls as they trundled off into the dark, and had mistaken them at first for the cries of screech owls.

Just as I was tiring of weeding, five shots rang out on the hillside quite close by.

# Chapter XXVIII

# "Call Me Any Time"

The following morning began like its recent predecessors—more beautiful than the one before, cool, sweet, filled with the songs of birds. But I knew from a glance at the sky that midday would bring crushing heat, causing the small trees to wither and the garden plants to droop.

Shortly after breakfast Alice trotted up to the house, carrying an object which, after only a little uncertainty, she deposited on her No. 1 Show treasure pile. It was the shinbone and hoof of a freshly-killed deer.

I tried to figure whether the fawns would be old enough to live without milk. Then I tried not to think about the matter because I felt pretty sure I knew who had shot the deer. But that evening, unable to continue ignoring my anger and unable to find a telephone number for the game warden, I called the sheriff's office.

Two deputies showed up just at dark and wrote copious notes as to what, where, and when. I was careful to betray no suspicions about *who*.

"The warden has two whole counties to patrol," one of them said. "But he'll be around to investigate as soon as he can make it."

I went to bed that night thinking I felt better for having reported the crime, like a citizen who had done her patriotic bounden duty. Later that night, however, I became violently sick. Drinking from the beautiful white waters of Oregon Creek at Easter may have given me intestinal parasites—or maybe the act of turning in a deerjacker had disagreed.

For many days afterward I lay in bed, zombie-like, unable to read or even to think. I watched house finches through my bedroom window, which even in the language of birdwatchers rates pretty low. I do not care for house finches; they are tweetling birds. Several factions were warring for the rights to a nest under the roof overhang, which was already occupied by a female and her brood. The redfaced father finch looked increasingly harrassed as he hustled bugs for this ravenous tribe. I saw him arrive with an insect in his mouth and hesitate near the nest where a stray adult female sat on a wire. She opened her beak. He, infatuated by just another pretty face, chucked his insect into it. That, I decided in my possibly mortal pain, was the ultimate meaning of "bird-brain." Maybe it was even the ultimate meaning of life.

Days followed days, a blur of gray sky, gray wind, the swamp drying up, dull mind, wanting to go out to water the dying trees, making lists of urgent tasks; stopping halfway, thinking this cannot matter, let it go. The time thick and tasteless as it dragged past and I longing for the weekend which would bring company, color, conversation, another point-of-view.

After what seemed years I was roused from my orgy of misery one morning by an indecently vigorous pounding on the front door. Who could be so rude as to sound so healthy? By what right?

"I'm Clay Dent, your game warden," announced the cheerful, grizzled grig, sweeping off his game warden's cap. "When did this deer-shooting occur?"

"I'm surprised the date isn't in the report," I said. "The officers wrote it down. About ten days ago."

I thought an expression of relief crossed his face.

"Well, it just got to my desk. That's about par for the course. I'll go out and have a look around."

A half-hour later he was back at the front door. He said, "It was X who shot the deer."

Although X had been my suspect, I was surprised by the speedy confirmation after so much time had passed.

"How can you be sure?"

"Found the spot up there on the hill where he shot the deer," said Warden Dent. "Then I followed the trail of dried blood down the draw. I asked myself, if it was me, where would I go through the fence down there by the creek? And I looked at about three likely places—great spots down there for hiding contraband, by the way—and I spotted the place where he carried the deer through the fence. Grass mashed down. Then I followed the trail right up to the old cabin. I guess he must have put papers down and butchered it on that."

"You could see all this?"

"You get used to knowing what to look for," he said modestly.

I was reminded of the Bedouin trackers of Saudi Arabia who, days after a sandstorm had whipped away faint tracks of a fugitive, could follow them across rocky terrain.

"Well," he said, "I can't get a search warrant now because ten days have elapsed."

"I wish there was a way of just scaring him before all the deer around here are wiped out."

"He has done it before and he'll do it again," the Warden said confidently. "They get careless. Like throwing the legs into the garbage where a dog could get at them, that was stupid. Next time—. You read detective stories? Then you know there's always some little trademark that gives 'em away."

He handed me his card.

"Well, call me any time. My working hours are any eight out of the twenty-four. D'you know what I was doing at 4 o'clock this morning?"

"I can't guess."

"It's the full o' the moon." He paused significantly.

I still declined to speculate although visions danced in my head.

"Hog poachers," he said. "There's this gang that trains Pitt bulldogs for rustling. A Pitt bulldog never makes a sound. And when the moon is full and bright, the rustlers don't need to use a flashlight except just for a few seconds at the last minute. The bulldog rushes in and holds down the pig while the rustler stabs its jugular vein with a sharp knife. No sound, except for one squeal."

"You must lead a dangerous life," I said sincerely. "This is wild country for one man to go up against armed criminals."

"Well, I'll tell ya. I've been in this game a long time and I know pretty well every move they'll make. Would you believe I'm sixty-seven years old? Now, sometimes you just get idiots blazing away at anything in sight, gun freaks that'll shoot the livestock in a rancher's pasture. But the men who rustle for profit are clever. Knowing how thin the law is spread, they usually stage several cattle rustlings on the same night, maybe fifty miles apart. Maybe they'll shoot the steer through the nostril into the brain, leaving no mark. High prices now for meat.

"Well, call me any time of the day or night if you have more trouble."

After he had gone I got to figuring and discovered that it had been only nine days since the shooting on the hillside, which made me smile. I spread the word around, casually, though, that the warden had been out checking on poachers and deerjackers. Shortly afterward, X and his van moved on to another area and all the shooting stopped. But there were then few deer left.

# Chapter XXIX

# The Great
# Demonstration Swim

On a midsummer morning that threatened to turn sweltering, I went to a neighbor's old orchard to pick apricots. An air of ancient loneliness hung about the place. The old farmhouse, the barn, and the people who had once lived there were all gone. A new tarred road curved through the orchard to a splendid house on a hilltop a mile away.

The fruit from the old trees was so ripe and heavy that it had begun to blanket the ground with a pinkish bloom over which the muddaubers buzzed. The apricots were small and very sweet. One tree had become entwined with an old, wild plum tree. I picked some of the tiny red plums to mix with the preserves for tartness, eating as I picked, and feeling goosepimples from the stillness of the place.

Later, sitting in the backyard at Meanwhile, I began pitting fruit, tossing the pits into a separate bowl. Alice again demonstrated her talent for figuring out what was up for grabs as against what she had not yet been offered, by helping herself to a pit which she carried off and played with. Then she came back and stood pointedly by the apricot bowl. I offered her a piece of fruit. New taste thrill. She came back about ten more times.

I mixed sugar and lemon juice as a syrup and poured it over shallow trays of the raw apricots, which I set on the picnic table in full sun. In a few hours the fruit would be stewed by solar power to a puffy, thick mass which could be poured into preserve jars and refrigerated. While the fruit was stewing, I became painfully absorbed in a cliffhanger being enacted in one of the tall sycamores overhanging the haul road.

A fledgling redshafted flicker had been left in the hole by parents who obviously considered it time for him to bail out. He was a large, strikingly handsome bird with a downy puzzled-looking head, glossy eyes, long beak, red markings on either side of the beak, and a black shield on his speckled breast. And he was paralyzed with fear. The suspense had been dragging on for two days. During this whole time the young bird had perched at the opening, extending a claw, retracting it, and screeching by turn indignantly and despairingly. The business was beginning to grate on my nerves.

Other birds were getting tired of the racket too. A flycatcher swept right up to the hole and uttered a rude cry. The flicker's parents flew past once and called to it—and that was the last he saw of them.

The temperature was well above 100 degrees. I decided to check the preserves and then go to the pond to swim. A honeybee had managed to wriggle under the plastic and get itself immersed in the jam. What a way to goo!

Fan Farkle came in from the field without her flea collar as I was leaving. She often lost it because of her smallish head. Usually Ferd would find it and bring it to me, playing with it until I noticed him. This time, though, Alice found the collar and brought it in. There was something tattling and nastily patronizing about the way they did this. I put the collar back on Fanny and spoke to her about taking care of nice things.

At the pond I floated for a long time, looking up at the pure blue sky. Every once in a while a gold minnow or a streamer of marsh grass would brush my leg. I was part of this cool green bowl with the cool blue lid. After I had crawled out to sunbathe, I felt a strong other presence. I glanced all around in the heat. Nothing. Then I bent over and peered under the board on which I was lying. There squatted a brown frog, its écru satin throat pulsing. The frog's golden eyes met mine but it held its ground.

Just then Kahn the Dalmatian pup arrived from another farm to play with Alice and a striking thing happened. Alice, after one unfortunate experience, had never swum. Once when a friend and I were in the pond, I had pulled the dog in and encouraged her to swim across the deep part beside me. Frightened, she broke away, swam to my friend and inadvertently scratched her. The friend uttered a cry of pain, whereupon Alice swam quickly away to the far side of the pond. A moment later I noticed her standing there among the reeds, her usually exuberant tail and her head hanging straight down into the water, literally immobilized with shame. I, the guilty party, got out of the water and called her to play on the bank. She responded at once and everything was all right again. From then on, though, she never ventured any farther than the shallows.

I could not have been more surprised on this occasion, therefore, when she suddenly plunged into the deepest part of the pond and began swimming quickly back and forth. While the pup and I goggled, she did about six graceful turns, her tail streaming out behind and a little smile on her face as she registered our astonishment. Just as she was about to call it quits, I cheered, which caused her to take one more turn. Then she bounded out, shook all over us, and dashed off into the field with her company. After this triumphant demonstration, however, she decided to rest on her laurels and has never since gone swimming. No law says a girl has to redeem her honor every day in the week.

Back at the house, Alice mock-pounced on Fanny Farkle and went through her Grand Guignol, Fanny protesting loudly as her head was disgorged and Ferd averting his eyes. I checked the thermometer in the shade of the back porch. It was 115 degrees. No wonder even the animals were acting like nuts. The preserves on the picnic table were cooked almost brown. I carried them indoors and poured them into jars. After finishing this sticky job and while fixing a cold drink for myself, I dropped some ice cubes into the animals' water bowl. Alice picked one out, perhaps thinking it some pleasant new kind of apricot, and carried it to the middle of the kitchen floor. She flopped down and, holding it awkwardly between her big paws, began sucking on it.

"Out," I said, pointing toward the open door.

She picked up the ice cube, rose arthritically, moved a few inches toward the door and again crashed down. I let her get away with it because it was too damned hot. Picking up my drink, I walked into the living room. The house—with all due respect to Sylvia Porter—was stifling. The only cool place was the bathtub but it was now occupied by Ferd, who was stretched out, long, thin, and apparently near death.

Suddenly it struck me that a most remarkable silence prevailed. I hurried out to the garden, crossed to the haul road, and peered up toward the nesting hole of the redshafted flickers. It was dark, empty, and silent. The old *For Rent* sign was up again. I sighed. It seemed to me that an echoing sigh came back from the woods.

# Chapter XXX

# Hottest Gun in the West

The drought continued. On my morning visit to the chicken house I detoured to the ravaged-looking garden to pull the last nubbly ears off the stalks of sweet corn. Banties like fresh sweet corn almost better than earwigs. Alice was pounding along the path ahead of me, but I paused for a moment to watch a black bumblebee gouging into the center of a royal purple Princess flower.

A Banty was just starting to rise on the nest. Then a warm egg stirred the straw. And the hen began to sing—not the coarse, triumphant cackle of ordinary hens but an actual high, happy, wavering, roof-of-the-mouth little paean of Baptist thanksgiving and praise. Virgil was right: these eggs were minor miracles.

Each afternoon now, a hard, cool wind blew for a few hours, threatening to tear the roof off the house. I noted on the calendar that the day was August 15. A memorable day. Exactly a year had passed. I sniffed the air and smelled autumn, and prayed for rain.

The level of the pond had dropped a foot in the past week and I was worried about the pressure in the well. In the entire year only one inch of rain had fallen!

Later that day I drove to San Solace to confer with the well-driller, but found the entire crew out on emergency calls around the county. Only a radio operator was present. The scene was unexpectedly dramatic.

"Field Post to Command Post," the radio blurted. "ZzzZZZzt! Come in—."

Burst of static.

"—Rancho San Pietro. Well lacks suction. We are presently trying to restore suction."

"Roger, Field Post, over."

I found myself worrying for the owners of Rancho San Pietro. One could if necessary haul household water from the nearest source, but to truck drinking water each day for hundreds of cattle was something else entirely. Quite aside from empathic concerns, cattle quickly lost weight when short on water. They became upset and gathered around the windmill, all bellowing at once in a panicky and frightening way.

But it wasn't just the water shortage and the ruined crops that worried local people. With the countryside tinder-dry, a single summer lightning storm could touch off a dozen forest fires at once.

When I managed to get the full attention of the Command Post I said, "I wish someone would come out and have a look at my pressure tank. I think I'm losing suction."

"Okay," Charley said. "Try to get someone there tomorrow."

I drove off in a mood of profound anxiety, stopping on the way home at the feedstore. The population of Banty roosters had become excessive, yet I had been unable to issue an invitation to a beheading. Even though Angelo had kindly offered to perform executions for me, it seemed to me that one was not a real farmer until one could face up to this sort of responsibility. I had read about a kind of miniature guillotine designed for beheading chickens.

I described it to Clarence at the feedstore and asked if he carried them. He looked puzzled.

"Nope. Never heard of anything but the axe. But I'll tell you how my mother used to do it. She always claimed that the meanest thing in the world was a danged old heifer. So when she had to kill a chicken, she would get the axe and the chicken and say, 'Stretch out your neck, you danged old heifer! WHAP! POW! ZAP!' And she said it helped her a lot."

"Then I would have to get a heifer first, wouldn't I? And learn to loathe her?"

"Yep."

"Well, thank your mother for me, Clarence."

He shook his head sadly.

"She passed on three years ago, Miz. Heifer kicked her in the head."

That evening the sky above the parched swamp turned blustery and gray.

The Countess Lillian, whom I had asked to dinner, drove in from San Francisco in her wire-wheeled 1929 Headley Flyer, wearing her chartreuse bombazine duster and the kind of scarf that did in Isadora Duncan.

Entering the driveway, she jammed on the brakes, started waving out of the window and shouting, "Tarantula! Tarantula crossing the drive."

Never having seen a tarantula before, I rushed out. There it was, fist-sized and hairy, looking surprisingly like the black rubber kind that practical jokers put in one's favorite chair. It seemed in an unholy rush to get to the other side of the road and crank up the rainmaking machinery. Seen in these circumstances, there was something positively endearing about it.

"Rain in seven days!" I exulted. "That's the best news in months."

"Let's hope your country folklore is reliable," the Countess said. "Now tell me, why do they cross the road?"

"Well, I don't know for sure but this is how I've worked it out. When they know it's rain-time they come up out of their holes in the low ground exactly a week ahead of time—which gives them a chance to find a hole somewhat higher up and drier for the winter. To get from low to high, you often have to cross the roads, and once they get there and into the new place, they make it rain. It just happens to take them seven days to do all this."

"Ah," said the Countess. "By the way, I've brought you a little book—."

She handed me a thin volume. *USDA 90463.* "Everything You've Ever Wanted to Know About Banties But Were Skeered to Ask."

I thanked her, my heart pounding, and opened it at once. The first sentence jolted me. In fact, that was as far as I ever got.

"Raising Banties," the author wrote, weeding out the weaklings in line one, "is not a Get-Rich-Quick-Scheme."

I must have turned pale. I know my jaw was waggling as I felt for a chair.

"Yes," said the Countess. "I thought you had best know. Soonest learned, least lost, as we used to say, or something equally tedious."

"I'll tell you one thing," I stammered. "It goes d-deeper than that with me and *Banties*. Making a million dollars isn't *everything* in the world. In fact, with inflation, it's even less."

"I couldn't agree with you more," she said. "Would you by any chance happen to have any little snacks made up?"

I apologized, pulled myself together, and invited her around to the back yard for cocktails.

"I'll have lemonade," said the Countess.

We had just begun to enjoy our drinks and the special quality of the hour when Alice started to chase something across the haul road. She hesitated and jumped backward. From the dense leaves on the far side

came an ominous buzzing that I had never heard before, too loud to
have been an insect.

"Rattlesnake!" exclaimed the Countess in a tone of authority. "Have
you a firearm on your person? A fowling-piece? A Saturday Night
Special? A gat?"

"No. No guns. Alice, come!"

The dog returned, looking puzzled, and I hurried her into the house.
When I returned, the Countess was headed toward her car.

"I carry a weapon in the glove compartment," she said. "Only an
antique dueling pistol. Takes a few minutes to load, what with ram-
ming in the ball. I've always thought I'd just use it as a bludgeon."

I returned to the roadside, some twenty feet from the source of the
buzzing. The Countess's lemonade glass sat on a nearby fencepost. I
placed my Martini carefully on another. Then I picked it up again and
took a strong sip, and returned the glass to the post.

The Countess returned, carrying a slim, beautiful pistol. I was re-
lieved to see that she also carried a small box of .22 bullets.

Her hands trembled as she tried to load the pistol and it took her
about ten minutes to get the first bullet in. It had to be loaded in a
rather peculiar way.

I said, "Don't forget to fan it."

Ignoring me, she extended her arm, steadied it, and blazed away.

The bushes quivered. The buzzing stopped. And then it resumed
again, more angrily than before.

"Here, let me," I pleaded.

The Countess later explained that her hand was trembling on the
first shot because she had known I was going to insist on shooting.

It also took me about ten minutes to load the little pistol and cock it.
Then, with a roar that clogged my ears, I let fly. The snake in the leaves
again stopped. It seemed to be listening in a bewildered way. I reloaded
the weapon, not much more quickly than before, and again fired,
directly at the point from which the menacing sound emanated. Again
a disconcerting pause. And again the rattling started in the same spot.

"I'm pretty sure I've got a bead on it now," I said.

I kept on blazing away, volley after volley about every five minutes.
Each time the snake stopped rattling. I would say, "Got him *that*
time!" But in a moment the buzzing would resume, madder than ever.

Between loading and firing, I had several sips of the Martini on the
fence post. If something didn't break this deadlock, I would have to
turn the dueling pistol over to the Countess and dash into the kitchen
for more courage. By this time my ears were deafened.

"CAN'T HEAR HIM. MUST HAVE GOT HIM THAT TIME!"

"That's because it seems to have stopped completely," the Countess said witheringly.

"Wonder if it's dead?"

"I *rawther* think he got bored with the whole business. Probably went home."

"Well," I bragged, "that's either one dead rattlesnake or it's one deaf one."

"As a matter of fact," the Countess said, "rattlesnakes are born deaf. They are sensitive to vibrations and other things."

I later read somewhere that they were also timid; but so was I when it came to rattlers.

In the kitchen I found poor Alice quivering from all the explosions. My trigger finger was black and badly powderburned. My ears still popped.

The Countess, sniffing around the refrigerator, exclaimed, "Oh, I *say. Cucumber sandwiches.*"

"I wonder what I ought to tell the neighbors when they ask about the shooting? They know I hate guns."

"Tell them," she said, "that you found out the truth about Banties."

That night I went to bed still feeling tense and only at a late hour fell asleep. At dawn I awoke to a strange, thrilling, but unmistakable sound—the patter of rain on the roof. Bless the tarantula, it had come through, and ahead of schedule. Suddenly I felt deeply rested and relaxed, even for once relatively at peace with the world. I lay there, trying to recapture the threads of bizarre but unimportant dreams through which the beautiful, miraculous rain kept streaming.

Few experiences in life can be so gratifying as the end of a long drought.

Chapter XXXI

# The 11,962nd Banty

One of the rules of writing about farming is that you are supposed to come up with at least one big bright idea that nobody else has thought of and be very cool about it. Your idea may soon be proved unworkable and be cancelled out entirely in the next edition, by which time, thank heaven, the reader should have moved along to something else. My big breakthrough came one morning when I noticed that the Banties had eaten all the grass in their pen, leaving bare, well-fertilized earth. If I moved the wire fencing to the other side of the chicken house, they would have fresh grazing. I could then plant my late cabbage family right there in a preweeded, precultivated, and prefertilized plot. Ruth Stout, I roared, eat your heart out! It was what we "scientists" speak of as an elegant solution.

And being a "scientist" at heart, I didn't just go ahead and move the fence to find out if rotating with Banties really worked. No. Instead I got out writing materials and began to compose a paper on the subject to present at the Oslo Congress of the International Banty Breeders Ladies Auxiliary. Scarcely had I hit on a scintillating title—"Rotating With Banties Management"—than there came an unseasonal cheeping at my kitchen door.

This was nothing new. The *only* trouble with Banties is that they get hysterical if you even *try* to reason with them about Zero Population Growth. Their fierce and insatiable drive toward motherhood cannot be daunted by any method I have thus far been able to figure out. If you allow a hen to set on a clutch of eggs in the chickenhouse, other hens keep pushing her out of the nest and sometimes succeed in taking over. The moment the chicks are born, at least half the hens in the chicken-house instantly go all cross and clucky with false pregnancy. If two hens have chickens at the same time, they start chick-snatching from each

173

other and the stronger winds up with most of the brood. At the same time half-a-dozen or so other hens will be reporting regularly for foster-aunt service, which means that they hang around the chicks, cluck over their Startena to encourage them to eat, and wind up consuming most of it themselves. Sometimes two or three hens will crowd into the same nest, adding fresh eggs to those with life in them. One hen may be the legitimate brooder on some fertile eggs. The second may have already hatched some chicks but elects to tuck them in at night on top of the brooding hen and her fertile eggs. The mother then climbs on top of this insane four-decker Banty sandwich, with tiny chicks' heads sticking out through her feathers.

If you try to thwart the incessant maternal urge by carefully collecting the fresh eggs every day, the hens go over, under, or through the fence and hide nests in the woods. If you clip their wings, as I have done, they learn to walk up the vertical wire mesh using the wings as balancing aids, and off they go to the wilds.

Those who succeed in nesting in the woods and their young are fantastic. Come sleet or snow, the moment the tiny creatures are out of the shell the mother hen proudly leads them home. I could never believe the strength contained in a quarter-ounce of day-old chick that had just made the Long March from a woodland nest. The mother of course immediately begins to drag them around the field, teaching a full eight-hour course in scratching. If rain begins to pour, she squats like an inflated mushroom with the chicks warm and dry beneath her.

The chicks are so incredibly healthy that only the few who are caught by predators fail to grow up. I once saw a mother Banty tackle a Cooper's hawk four times her size that had fastened its talons in a chick. Alice and I arrived at the same time and the huge hawk fled, leaving an injured chick that grew up strong and healthy.

Thus, on the day when I had settled down to compose my paper only to hear an unseasonal cheeping at the door, I knew what to expect.

A black and white Banty displayed eleven tiny yellow and brown young. By my latest count this made 11,959 Banties. Something had to be done. The only way I knew to stamp out fertile eggs was what I had been avoiding all along—the beheading of roosters.

Next morning early, with hardened heart, I held the axe poised above the chopping block. In the other hand I held the legs of the scheduled victim, a beautiful rooster whose clear, caramel-bright eyes regarded me in gentle questioning. I raised the axe higher. Suddenly I heard a burst of high, devotional singing. And around the corner of the house, stepping with affected care so as not to snag her skirts, glancing this way and that as if prepared to at least *consider* the petitions of all sinners in her line of march, and singing her little heart out, came one of the tiny speckled hens from the Original Baptist brood.

In one horrifying appraisal she took in my posture, the axe, the rooster. I froze with the enormity of my intent, but she went right on singing, chilling my soul. I felt my fingers loosening on the axe. I felt the hand that gripped the rooster's legs opening.

The axe lowered. The rooster, uttering a surprised "Clur-rrk!" flapped to the ground. He shook down his feathers. And just to prove that nobody has much use for an altruist, he chased off in hot pursuit of his deliverer. He caught up with her. She obligingly squatted, without interrupting her hymn.

I shrugged with self-disgust, picked up my axe, and went indoors. I phoned some neighbors who had once carelessly evinced interest in getting Banties to solve their insect problem. By dint of persuasion bordering on blackmail, I got them to say they would accept the rooster and the latest hen with chicks.

But, good grief, that still left 11,900-plus. Certainly too many to have to buy feed for in the winter when they did not lay much.

In my time of desperation, however, another neighbor phoned to ask whether she could borrow two broodies to hatch some large eggs. This gave me a bright idea that led to my advertising the Dial-a-Broody-Banty Service in our farm shopping news. It was an immediate success and probably will pay my expenses to the Oslo Congress.

There ensued other ups-and-downs with the Banties—but I shall be saving it for a paper on pseudocyesis that has been requested by the Banty Husbandry Management Convention.

Some of my Banties, I might disclose, started eating their own eggs. I quickly consulted the county agent, who declared it "an extremely difficult behavior pattern to break," and who promised to write to a distant farm advisor who was really into it. In good time I received a letter from one H. Bill Murchison, who suggested a number of homey ideas, including making curtains for the Banties' nests. The letter closed with, "However, the surest way is to have a chicken dinner." *A* chicken *dinner?*

On an impulse I consulted Dr. Frank Miller's "Wonderful World of Animals" to see if he could shed any light on the problem of Banty egg-eaters.

He replied heartily: "Yes, Mr. Chips your canary could indeed be jealous of your new boyfriend. Feather plucking and frustration are often related. . . . Whether he needs cuddling, a counselor's couch, or a specific physical diagnosis (dermatological or otherwise) can't, unfortunately, be determined from this distance."

Below that in his newspaper column was a letter from a woman who wondered if her dog was drinking too much. The paper dropped from my nerveless fingers.

If there is one thing I have learned it is to accept the fact that some problems may have no solutions. Banties may fall in this category.

Chapter XXXII

# Country Women, Traditional Man

Barbara, having developed many personal interests, has been able to convert a hobby, the one thing she most enjoys doing, into a useful and rewarding new field of work. She buys small old houses and remodels them into attractive, low-cost dwellings. In our area this particular type of housing is greatly needed. Usually the small-home buyer has little recourse except to tract developments.

When Barbara arrived at the farm she had already taught herself a number of practical skills, but we both felt a need to learn more.

For some time we had been receiving *Country Women* magazine, published by a collective of women in Mendocino County from editorial headquarters in a cabin room ten-feet-by-twelve. The prose and poetry seemed exceptionally fine and the how-to-do-it articles offered the first reliable information I had been able to find on how to understand your well-pump-pressure-tank relationships, how to put out forest fires (first you tell the nearest man to go make some coffee and not to forget the doughnuts), how to raise, educate, medicate, understand, and enjoy your children, how to attain harmony with nature through acceptance of the cycles of birth, death, and rebirth, how to dehorn your goats, and much, much more. When we received an announcement of the workshops they would be teaching at the Third Annual Country Women's Festival up among the redwoods, it seemed like a good idea to go.

I am pleased to report that we found the independent spirit running strong at this foregathering of feminists from up and down the West Coast and from states all across the country. In a rural setting, it is possible for a woman to live almost free of man's unerring imprint, the profit system, and of male corporate dominance, *if* she has taken the trouble to develop competence in the daily details of survival. These women were reaching back into the history of matriarchy to the time when women developed agriculture, potting, weaving, food preservation, the use of herbal remedies, home-building, and the decorative arts. Everything they were doing represented work or art forms which came natural to the female, but which patriarchy had denied them. There had been no forums for their music, medicine, architecture, decorating, and little for their painting, writing, and drama. Even in agriculture, they were the put-down sex, as a reading of that popular magazine *Mother Earth News* (acronym MEN), with its ploughboy interviews and emphasis on a land movement dominated by men, quickly conveys.

It was hard to make choices among the Festival expertise—workshops in carpentry, electricity, plumbing, self-defense, chainsaw *management*, wood-splitting, shake-making, gardening, dance, total orgasm, auto mechanics, stained glass, goats, journal writing, ageing—an endless feast, it seemed, of womanflow knowhow. (My only quarrel with the literary self-discovery of the female as currently practiced is an infatuation with such unfortunately contrived words—childcontent, womanspirit, personhood, heartflutter, earth-child-self, dark-inward-hidden-self, worldflow, psychicspace, and more. Which is to niggle.)

As I read so many of these new publications I often think, "It would be nice to have a man's point-of-view on this." And then it hits me: but it's just the reverse of such sexism that I have been feeling all my life in the male-dominated media, making me feel an alien in my own land since birth. When has a woman ever been able to identify with the point-of-view portrayed in most publications, films, popular music, spectator sports, television programs, and so forth? Now women are creating their own magazines, newspapers, publishing houses, pouring forth products for a market that has always hungered for them.

Many of the little magazines were apolitical, but because they represented a burgeoning social force and an economic counterweight and filled a growing cultural need, they were intrinsically political. Simply by withdrawing womanenergy from the patriarchy, they were hotly political.

About these young country women, I noticed, there was not the self-pity that one sometimes noted among the middle-class unisex rebels of the sixties and seventies who, one way or another, melted back into the mainstream culture while cursing the conspiracy of their parents and their births.

Countrywomen were not, however, above sartorial vanity and cattiness. I overheard, while trying to figure out my workshop schedule, a girl wearing Levis and a Madras bedspread say to her companion, in the manner of a Coco Chanel slumming in a boutique—as both of them sized up a third countrywoman: "That's the first time I've seen *corduroy* cut-offs slit up the front!"

Quickly realizing that her comment was unbecoming of sisterhood, however, she added, *"Interesting."*

Barbara and I, an island of fiscal responsibility and squareness in a sea of Don't Care, indulged in a particular kind of conversation we resort to when we are really trying to eavesdrop on what everyone else is saying or doing.

"I have heard," I said, "that Science has been able to achieve parthenogenesis with Cornish Rock hens."

"Science," she replied witheringly, "has always been confused by finding little itsybitsy baby teeth turning up inside of the ovaries of unpregnant virgins. Face the facts. If you can have premarital sex, why can't you have presexual denture-wearers?"

"It sounds reasonable to me. If jugs have ears, why in the name of a pluralistic society shouldn't ovaries have teeth?"

Two girls standing near us were saying that umbilical cords were not being cut nor even bitten off any longer.

"We just let 'em dry up and fall off," one said.

Barbara and I joined a circle around a shade-tree mechanic named Sandy who was teaching the repair of both foreign and American cars from the rear of her VW van. She was a cool twenty-year-old resembling a junior Julia Child, whose confident manner and voice radiated a natural ability to teach. As she spoke she passed around parts and tools, insisting that her students, squatting in the dust, feel them, test their tension if any, and stick their heads into the engine. This was smart of her. Women are scared of the sexual/technical mystique of tools. Sandy freely admitted that the theory behind what she did to repair an engine eluded her.

"That's the difference between a two-week course in dirt mechanics and a two-year community college course in automotive engineering," she said glibly. "Helps to keep the colleges going."

She showed us how to fix a "well-gapped sparkplug" by measuring between 2500ths and 2800ths of an inch with a dandy little fine-bladed tool. I felt quite pleased, considering that my past linear minimum record of *anything* had veered from, Oh! about half-an-inch on up to three inches; and put it down to a triumph of sheer tenacity rather than, for example, womanflow.

Obviously we would not learn enough in two hours to repair our cars of any grave disorders. But the next time a gas station mechanic warned us that we dare not drive another ten miles with that frayed fan-belt, we might detect that he was actually pointing to a little fan-belt that kept the air conditioner going for a few hours each summer. We were learning not to be ripped off, hoodwinked, or flimflammed, and possibly, how to drain the oil and clean the oil filter. If we were quick studies, we might be learning how to avoid a $75 tune-up by installing new points. The most amazing if perhaps not the most useful thing that I learned about points was that they is singular; there are no *plural* points.

"More and more women are getting into working in the parts shops," Sandy said. "A Porsche engine looks like two VW van engines of 37 hp. And don't forget to lubricate your cam shaft. There are all kinds of horrifying little ways in which you can blow your engine. The reason a VW farts all the way down hill is because it has a measured exhaust."

The sun had climbed almost to treetop level and was beginning to melt the chill, its warmth filtering pleasantly upon us. A Grecian-looking girl with dark curls and brown body ripped off her purple *Sappho Lives* T-Shirt to disclose beneath it a white *Superwoman* T-shirt and passed her can of Coors among us. Then the final layers of shirts and sweaters started coming off.

Sandy, perhaps feeling that she might be losing some of us to thoughts of lunch, grabbed our attention with a crack about the Enemy.

"I learned auto mechanics on a commune from a man. And every time I leaned into the back of the VW to work on the engine, he tried to goose me. Which is just one of the rip-offs you get into, unless you can do it yourself."

My attention had begun to wander, what with the intellectual drain of alien matters. More out of genuine concern than a fondness for alliteration, I asked a young woman sprawled nearby in only her bib overalls whether the bib was not uncomfortable on bare boobs. She said, "No, it's a lot more comfortable, actually, to have the weight on your shoulders instead of around your waist," and proceeded to show me the added bonus of all her great pockets. A tiny, perfectly shaped blonde baby wandered into our ambit, peed in the dust, was picked up by her mother, who wore patchwork coveralls over nutbrown skin, and presented with a nipple. I began to understand that the new country "dress" was above all else functional.

"I have a horror story," Sandy was saying, as I wandered off, "about a distributor locknut screw on my Pinto—."

Terry, who had close-clipped blonde hair and rather reckless brown eyes, was conducting a carny spiel on Plumbing under a nearby redwood, waving odd bits of pipe in the air. I could not resist joining an obviously galvanized class.

It was quickly clear that Terry was out to reduce the plumbers' guild and orthodoxy to blubbering jelly. While waving the hunks of pipe around, she referred to them disarmingly as "like—take a *you*-know." But when she knew one of these sections by name, she really let you hear it—unless she disdained to do so whimsically because it was sexist in origin.

With hand aloft, dramatically: "Now, this is a *union!* You can use it to connect any kind of Rube Goldberg construction of pipes, as we have done at home (and we have *plenty* of water pressure between our two houses). You can increase or decrease your pipe size. EASIEST THING IN THE WORLD! They'll tell you it's *hard!* (I never can remember which are male threads and which female so I call them inside and outside threads.)

"And *this* is a COUPLING. You know, you take a thing and—you know—and you attach it to another thing. *Anything* fits into *anything.* They'll tell you it *doesn't!* I REPEAT: you can connect *anything* to *anything.*"

Although I was not yet sure if Terry knew a PVC from a T-joint, her delivery was certainly more refreshing than that of the normal run of male plumbers who had patronized me and my own domestic drainage crises. ("What you want to do with your solids, little lady, is you want to hold your *flushing mechanism* down a *little longer.*")

"—and don't let them tell you you can't!" blonde Terry iterated. "All you have to do is decide how to do it. I go into plumbing supply houses and just play around with the different kinds of pipe until I figure out my problem. They get used to you—."

"How can I wire my cookstove to a hot-water heater?" a girl asked.

Terry waved two hunks of pipe. "Well, first you take this thing and you know and you attach it to *this* thing'. '

"But I got a shock," the girl said.

An intellectual looking young woman spoke up. "There's something called electrolysis—." And, speaking quietly and crisply, she proceeded to recite what sounded like the entire chapter of a text book on electrolysis as engraved on a photographic memory plate.

Terry said, "Yeah, they talk back and forth in these classes about new kinds of pipe."

There followed a rowdy bit of input from the audience on the subject of plumbing merchandise sold by a certain mail order firm and someone said, "You can use plastic PVC four-inch pipe for sewer piping too."

As Terry was repeating that you could adapt anything to anything and "don't let *them* kid you!" a camper truck with a little shingled roof was driven past by a man who was staring as nearly straight ahead as one could do who had to navigate around trees. We heard bleating.

"The sheep!" people started saying. "The sheep have arrived."

The sheep, which turned out to consist of a single, enormous ewe, broke up Terry's assault on Plumbing, except for a hard core who wanted to handle the pipe wrenches.

In Sheepshearing, all the naked children were crowding around, asking, "When are you guys gonna shave her?"

The question was directed mainly to a dark, slim girl from the nearby countryside, who sat with the huge, fat ewe pulled backward onto her lap. Then she began working slowly, carefully, with her manual cippers, layering away the thick fleece. The ewe's pale, dull eyes rolled upward from time to time, but otherwise she seemed to have been tranquilized.

Everyone who came along asked, "Why doesn't she try to get up?"

The shearer said, somewhat sheepishly, "Because sheep are too dumb to turn over."

Later I watched a show on television in which it was reported that the sheep's fleece is often so heavy that the animal cannot turn over. When a sheep falls on its back, a smart sheepdog tugs it upright again.

Children paused to plunge their fingers irresistibly into the fleece and then moved them upward as if to test the texture of the stupidly rolling eyes. No one pulled the children back. Countrywomen interfere with what their children are doing only 1) if the child may be hurt or 2) if another creature may be hurt by it.

The girl shearing the sheep had beautifully shaped hands, as everyone observed; but she must have been suffering excruciatingly, buried in the hot sun beneath that living mass of wool as her fingers patiently clipped. The slow care with which she worked astounded me. At county fairs 4-H boys, using electrical shears, compete for prizes, with little concern for the blood where the sheep are nicked. In fact, as might be expected, there is a world's record for sheepshearing, held by an Australian who has clipped four hundred sheep in eight hours at one dollar per sheep. This girl, all lamb-gentle womanspirit, would win no prizes and none was offered.

Down the hill beneath trees, a large acoustical music group was playing softly with a nice beat, with flutes, drums, dulcimer, and tambourines, learning each other's styles. The musicians were proud of improvising their own woman-energizing, liquid, sweet, mystical, screw-the-musicians-union arrangements.

Late in the afternoon I returned from an icy swim upstream to find women sprawling in every possible stage of happy déshabille—a

Breughel composition with no codpieces—around the focal point of the festival, the entrance to the dining hall. Countrywomen were leaning against tree trunks, writing, sketching, rapping, reading, strumming musical instruments, and—in another part of the forest—learning to resist rape. A black girl was stroking the long, prematurely gray hair of her white sister, both of them taking pride in the boldness of the wearer of long gray tresses. Off in the woods at intervals yelling erupted from women in the self-defense workshop. That morning I had seen a note on the bulletin board: "Help! Canvas needed for punching and kicking dummy." It sounded as if they had found the props. As often during the festival, I found myself wondering whether the traditional *macho* had the merest clue of what lay in store for him. How would he react the day it all came down? Break out the National Guard? Drop the Bomb? Restore witchburning to respectability?

As the dinner hour approached, a heartstopping rumor spread that the menu would feature, if not spotlight, vegetable lasagna. Almost every meal thus far had consisted of ingenious new combinations of squashes floating in their own limitless juice.

A bare-bottomed toddler named Oedipus had been wandering around, sucking on an object finally identified to me as a speculum— which, certainly, nobody was about to take away from him. Now his six-year-old sister Eurydice rushed up and pointed him in the direction of the forest gloom.

"Go tell them it's *vegetable lasagna*," she said, "and they're to come at once. *Vegetable lasagna!* Can you say that? And hurry!"

He looked at her dubiously. And then, tugging at the place where he sensed the memory of a diaper and still sucking at his diagnostic tool, he toddled off into the twilight.

When at last the whole mob had squeezed into the dining hall, it was announced—to groans—that only a single small serving of lasagna could be permitted each person because the day had brought such an unscheduled pouring in of country sisters. But as usual there was good sopping bread and milk and a gulp of red wine to add a festive tingle, and excellent spirits.

An English girl, Tinley, described to us how she lived with her son Noah in a giant redwood stump on a mountaintop commune. She looked in glowing health and together and as confident as one could be. Her only problem, she said, was that her parents back in England simply could not accept the fact that she was building a life and a future for her infant and herself in a California tree stump. She worried because *they* worried.

"They cannot understand that we have everything we could possibly want or need and that we are living in the finest way we can imagine," she said, "and that we simply could not be happier. Of course, we'll have to *weatherproof* the stump a little for winter. But we have plenty of food and fresh air. If we need clothing, the commune gets it at the Good Will and distributes it. Noah has other children to play with and we're hoping to start a private school soon. But my parents can't seem to understand how all this can be."

Tatia, another bright, healthy specimen, said she had come west from Woodstock with her daughter Luck. And they too were finding the rural life—which in her case involved people living in separate homes on a cooperative spread—rewarding. Some of their group received food stamps, some earned money with crafts, or held jobs; and occasionally they worked out deals with growers in the Central Valley to get truckloads of tomatoes and fruit, which they canned for winter (and which must have involved some sacrifice of conscience, considering the liberal use of sprays on such produce).

Juliet, a charming, dark-eyed woman with a thirteen-year-old daughter, lived on a different type of commune that had been harder to find. She played the dulcimer and composed her own music. The piece she played for me in my cabin one afternoon had been improvised to celebrate the unscheduled birth of a baby the first night of the festival—a romantic, tender, almost gypsy-wild composition. They lived with other artists in a commune planned for the ultimate in privacy. It was good to think of her seated on a stump in a mountain meadow in early morning, her fingers flying over the dulcimer, and of the others who had beautiful surroundings for their art.

Some "country women" currently lived in cities and attended skills workshops, hoping eventually to break down sexist barriers to the trade unions. Mary, for example, drove a Yellow Cab in San Francisco at night, worked as a contractor's assistant by day, (which required no Carpenters Union card) and in her *spare* time was organizing a class action suit.

"A funny thing happened on my first carpentering job," she said. "I was taking karate at the same time, because it's a useful thing to know if you happen to drive a cab at night. I had this summer cabin on a hillside all jacked up and, after a great deal of study and work I had managed to replace the rotten floor. The day it was finished, I felt great. I removed all the jacks except the one on the downhill side. And then I impulsively gave a karate yell and whacked it with my hand. The cabin rolled downhill."

Toward the end of the meal, with all going amicably, one of the organizers shouted for attention. Climbing onto a chair, she announced to a stunned audience that the four-piece rock band engaged for the dance that night would not be available.

"We withdrew the invitation," she shouted, "because their lead guitarist is a *man*."

Howls, groans, screams of anguish.

"We felt," she hurried on, "that if even one of our sisters objected to having a single man at the festival, her rights should be respected."

Even louder howls, groans, and screams, but this time with the addition of applause and whistles. Someone called for a voice vote on the decision. It was the only time I could remember ever hearing a proposal for a voice vote shouted down. Things looked sticky indeed for the chairperson. Womanpride was at stake but so was the prospect of womanpleasure.

Although many good musicians were present they were not rock players. Finally a hand-count was taken upholding the decision to forget about the guest artists. And in a split-second table conversations resumed. The mood of discord reverted to one of sisterhood, sweetness, and light.

Next day on the drive home, Barbara and I compared notes. We agreed that it wasn't so much what we had learned as the new confidence we felt, the new womanparameters. The thing I had most enjoyed was learning the simple knack of splitting redwood shakes with a frow and a wooden mallet.

The younger generation of country women had impressed us. That was wood-butcher country. Even so, we had been surprised to find that almost every woman present, somewhere along the line since puberty, had taken part in some aspect of shelter-building. Nothing seemed to faze them. I found myself wondering again about the reaction of traditional man when he found out where the revolution had gone. Or was he just pretending not to know?

I tried it on William in San Solace who may, I believe be regarded as traditional man. I told him about the country women voting not to have a rock group in because their lead guitarist was a man. He stopped dead in his tracks and gave me his slapped look. On such occasions there is a reflex silence on his part followed by a back-up silence representing his calculated decision that it would be ridiculous to respond to such bias.

After I had heard both of these silences, I quickly pointed out to him that women musicians had always been excluded from important roles in mens' groups and from serious consideration of their compositions for symphony concerts and that outstandingly talented women directors had been deprived of the right to direct major orchestras—until, just recently, *one*.

"These country women," I said, "were trying to prove to themselves that they could make their own, good music—which they did."

"Yeah, I dig that," William said too quickly as always, meaning that the whole subject was hopelessly traumatic. Somehow the concept of simple *equality* could not be other than threatening to him personally. In every other respect he was a fair-minded man so I always kept hoping.

I wonder from time to time what sort of Meanwhile Barbara and I are creating. It is certainly a curious one in some respects. Who ever heard of growing cranberries in a bog, as I am doing, not more than fifty feet from grapes, which require good drainage? They are not companion crops that the county agent would ever have suggested. Nor pussley next to henbit, unless I miss my guess. Yet I am reassured by Robert Rodale's assertion that a good farmer always experiments and takes risks. If trial and (especially) error were the only criteria, we should be the greatest farmers since the world began.

It is sometimes comforting to reflect that one can always pass the buck to Nature. At present, according to meteorologists, we are coasting into a new Ice Age that may markedly chill the food production process within a decade. It would be good to know whether God has a sign on His desk saying, "The buck stops here." Maybe we'll postpone our solar-heating system for the roof and convert the swamp to algae-pond hydrogen fuel and methane gas production. Maybe the IRS will grant us an "involuntary conversion"—.

So—what if you build your Meanwhile and it doesn't turn out exactly as originally conceived? So what if you have to compromise? Stop dreaming, maybe, of becoming a Condor Station, or don't win First Prize in the Junior Marrow category *this year?*

To a Meanwhiler, serendipity is lifeblood. Inch-worms are milestones. Mulch is marvelous.